MW01254646

Foundation Engineering

About the Author

Richard L. Handy is a Distinguished Professor Emeritus in the Department of Civil, Construction and Environmental Engineering at Iowa State University. A sought-after teacher, he served as the major professor for over 60 graduate students, many of whom have gone on to make major contributions in geotechnical engineering. A large number of former students and associates recently collaborated to endow a Professorship in his name, and a book of collected papers was issued in his honor.

Dr. Handy may be best known as the inventor of Borehole Shear Tests that perform in-situ measurements of cohesion and friction in soils and rocks. The soil test was used in snow when he and six engineering students were conducting research on an epic voyage of a large ship in the ice-bound Northwest Passage. They also observed the catenary shape of an igloo, which he later adapted to solve a problem that had intrigued Terzaghi, to mathematically define arching action in soils. The analysis revealed that conventional analyses are on the unsafe side and explained a wall failure where there were four fatalities. It received the Thomas A. Middlebrooks Award of the American Society of Civil Engineers.

Dr. Handy also was active in geology. He proposed a variable-wind hypothesis to explain the distribution of wind-blown silt (loess), and showed that the rate of growth of a river meander slows down in time according to a *first-order rate equation*. He then applied the same equation to rates of primary and secondary consolidation in engineering. In recognition of his contributions to geology he was elected a Fellow in the Geological Society of America and the American Association for the Advancement of Science.

Known for his sense of humor, Dr. Handy liked to point out that it is better to have a joke that turns out to be an invention than an invention that turns out to be a joke. His *The Day the House Fell*, published by the American Society of Civil Engineers, Reston, VA, for non-engineers, became a best-seller. His book *FORE and the Future of Practically Everything* published by Moonshine Cove Publishing, Abbeville, SC, adapts first-order rate equations to practically everything, including track world records and baseball home runs.

Dr. Handy also founded and is the Past President of a company that bears his name. The company manufactures and sells geotechnical instruments, with emphasis on in-situ test methods that were created and developed under his direction.

Foundation Engineering

Geotechnical Principles and Practical Applications

By Richard L. Handy, Ph.D.

Distinguished Professor Emeritus
Iowa State University

New York Chicago San Francisco
Athens London Madrid
Mexico City Milan New Delhi
Singapore Sydney Toronto

Library of Congress Control Number: 2019944581

Foundation Engineering: Geotechnical Principles and Practical Applications

1 2 3 4 5 6 7 8 9 LCR 23 22 21 20

ISBN 978-1-260-02603-0
MHID 1-260-02603-5

The pages within this book were printed on acid-free paper.

Sponsoring Editor
Lauren Poplawski

Editorial Supervisor
Donna M. Martone

Acquisitions Coordinator
Elizabeth Houde

Project Managers
Poonam Bisht and Jyoti Shaw, MPS Limited

Copy Editor
Mohammad Taiyab Khan, MPS Limited

Proofreader
Alekha C. Jena, MPS Limited

Indexer
Jerry Ralya

Production Supervisor
Pamela A. Pelton

Composition
MPS Limited

Art Director, Cover
Jeff Weeks

COVER PHOTO shows construction of a Rammed Aggregate Pier® that substitutes coarse aggregate for concrete and affects soil around and between piers. Photo is by Alex Schick and courtesy of Peterson Contractors, Inc., Reinbeck, IA.

Contents

Preface ... xv
Introduction ... xvii

1 Defining What Is There ... 1
- 1.1 The Three Most Common Construction Materials ... 1
- 1.2 Two Classes of Foundations ... 2
  - Support of Deep Foundations ... 2
  - Expansive Clays Can Be Expensive Clays ... 2
  - End Bearing on Rock ... 2
  - Ground Improvement ... 3
- 1.3 Residual Soils ... 3
  - Travel Is Wearing ... 3
- 1.4 Soil Layers Created by Weathering ... 4
  - Topsoil "A Horizon" ... 4
  - Subsoil "B Horizon" ... 5
  - Shrinkage Cracks and Blocky Structure in Expansive Clays ... 5
- 1.5 Vertical Mixing in Expansive Clay ... 6
- 1.6 Influence from a Groundwater Table (or Tables) ... 6
  - Groundwater Table and Soil Color ... 6
  - A Perched Groundwater Table ... 6
- 1.7 Intermittent Recycling ... 7
- 1.8 Soil Types and Foundations ... 7
  - Influence of a Groundwater Table ... 8
  - Pull-up of Deep Foundations by Expansive Clay ... 9
- 1.9 Agricultural Soil Maps ... 9
  - The Soil Series ... 9
- 1.10 Distinguishing between Alluvial Soils ... 9
  - Rivers and Continental Glaciation ... 10
  - Meanders and Cutoffs ... 10
  - Oxbow Lake Clay ... 11
  - Alluvial Fans ... 12
  - Natural Levees ... 12
  - Slack-Water (Backswamp) Floodplain Deposits ... 12
  - Air Photo Interpretation ... 12
- 1.11 Wind-Deposited Soils ... 13
  - Sand Dunes ... 13
  - Eolian Silt Deposits ... 13
- 1.12 Landslides ... 14
  - Landslide Scarps ... 14
  - A No-No! Landslide Repair Method ... 15
  - When Landslides Stop ... 16

	Recognizing Landslides	16
	Not a Good Place for a Patio	16
1.13	Stopping a Landslide	16
	Drainage	16
	Structural Restraints: Piles, Stone Columns, and Retaining Walls	17
	Chemical Stabilization	17
	Drilled Quicklime	17
1.14	Rock That Isn't There	18
	Near-Surface Features	18
	Shallow Caverns and Sinks	19
	Locating Underground Caverns	20
	Abandoned Mine Shafts and Tunnels	20
	Tunneling Machines and the Rock That Isn't There	20
1.15	The Big Picture	21
	Mountain Ranges, Volcanoes, and Earthquakes	21
	Soil Responses to Earthquakes	21
	Earthquake Recurrence Intervals	22
1.16	The Walkabout	23
	Problems	23
	Further Reading	24
2	**Getting along with Classification**	**25**
2.1	A Hands-On Experience	25
2.2	An Engineered Soil Moisture Content	25
2.3	Standardizing the Plastic Limit Test	26
	The Plastic Limit in Engineering	26
2.4	Going from Plastic and Remoldable to Liquid and Flowable	27
	Standardizing the LL Test	27
	The Fall Cone Test	27
2.5	The Plasticity Index	28
2.6	Atterberg Limits in Soil Classification	29
2.7	WWII and New Rules for Soil Classification	30
2.8	Atterberg Limits and Criteria for Expansion	31
2.9	Kinds of Clay Minerals	31
	A Layered Crystal Structure	31
	An Expansive Crystal Structure	32
	Going Tribal	34
	When Sodium, Na^+, Replaces Calcium, Ca^{++}	34
	Drilling Mud	34
2.10	A Hands-On Test for Expansive Clay	34
	Field Test	34
2.11	Some Clues to Expansive Clay	34
2.12	Measuring Soil Particle Sizes	35
	Statistical Interpretation	35
	Defining Clay Size	36

References 57
Further Reading 57

4 Soils Behaving Badly 59
4.1 Expansive Clays 59
Expansive Clay in a Consolidation Test 59
4.2 Two Classes of Expansive Clays 60
Type G Clays 60
Type P Clays 60
How a Layer of Expansive Clay Can Cause Trouble 60
Nature's Color Coding 60
4.3 Sorting Out Floodplain Clays 61
What Makes River Floodplains Wide 61
Braided Rivers 61
Meandering Rivers 61
A Shift from Braided to Meandering 61
4.4 Floodplain Soils of Meandering Rivers 62
Oxbow Lake Clay 62
Depth and Shape of an Oxbow 62
Slack-Water or Backswamp Deposits 62
4.5 Deep Tropical Weathering and Expansive Clay 63
4.6 A Guide to Expansive Clay 63
Crystal Structure in Control 63
4.7 Field Evidence for Expansive Clay 64
More Bad Karma 64
4.8 Managing Expansive Clay 65
The Chainsaw Method 65
Structural Slabs, Grade Beams, and Piles 65
Stripping off the Active Layer 65
Observations of Strange Field Behavior 67
4.9 The Replacement Method 67
How Does It Work? 67
New Rule for Control of Expansive Clay 68
Clues to Between-Layer Stacking of Water Molecules 68
Hypothesis 69
Why Does Clay Expansion Stop at 3 Layers? 69
What's in a Name? 69
4.10 Chemical Stabilization of Expansive Clay with Lime 69
4.11 *Collapsible* Soils 70
Delayed Collapse 70
Collapsible Alluvium 71
4.12 Regional Changes in Properties of Wind-Deposited Soils 71
4.13 Quick Clays! 72
Vane Shear Does Not Just Measure Soil Cohesion 72
4.14 Liquefaction! 73
Identifying Vulnerable Soils 73

	Earthquakes, Volcanoes, and the "Ring of Fire"	73
	Made Earthquakes	74
4.15	Pretreatment to Prevent Liquefaction	74
4.16	Earthquake Dynamics	75
	Recurrence Intervals	75
4.17	Quicksand	76
4.18	Blessed Are the Computers But Will They Really Inherit the Earth?	76
	Problems	76
	References	77
5	**Stresses in Soils**	**79**
5.1	Concentrated Stresses	79
5.2	Adapting Boussinesq Theory	80
5.3	A Snag in the Relationship	81
5.4	Approximating the Pressure Distributions	81
5.5	Preloading	82
5.6	A Plate Bearing Test as a Model Foundation	84
5.7	Performing a Plate Bearing Test	85
5.8	The Progressive Nature of a Bearing Capacity Failure	86
5.9	Plate Bearing Tests on Weathered Soil Profiles	86
5.10	Foundation Stresses Transferred to Nearby Unyielding Walls	88
5.11	Strength Gains from Aging	89
	Interruptions during Pile Driving	90
5.12	A Convenient Maximum Depth for Pressure Calculations	90
	Problems	90
	References	91
6	**Evaluating Soil Shear Strength**	**93**
6.1	Bearing Capacity and Settlement	93
6.2	Friction	93
	Friction Angle and Slope Angle	94
	Amontons' Second Law	94
	The Greek Connection	95
	Coulomb's Equation	95
6.3	Friction Angle in Soils	96
	Dilatancy in Design	96
6.4	A Direct Shear Test	96
	Influence of Layering	97
	The Borehole Shear Test (BST)	98
6.5	Unconfined Compression Test	100
6.6	Mohr's Theory	100
	Pore Water Pressure	102
6.7	A Difficult Problem	102
	Stage Testing	103
	Lambe's Stress Path Method	103
	What about the Intermediate Principal Stress?	103

6.8 Statistical Analysis of Test Data 104
R^2 (*R* squared) 104
Triaxial Shear Tests 104
Problems 104
References 105
Further Reading 105

7 Shallow Foundation Bearing Capacity 107
7.1 Bearing Capacity versus Settlement 107
Temporary Excess Pore Water Pressure 107
Unanticipated Loading 107
7.2 Fair Warning 108
Two Kinds of Decrease in Pore Water Pressure 108
Drainage 108
Sensitive Soils 108
7.3 Foundations on Compacted Soil Fill 108
Procedure and Performance Tests 109
Cut-and-Fill 109
7.4 Bearing Capacity Equations 109
Equation Development 109
7.5 Prandtl-Terzaghi Analysis 110
Rough Base, Smooth Base 112
Meyerhof's Modification 112
7.6 Terzaghi Bearing Capacity Factors 112
Local Shear 115
Alternative Solutions 115
7.7 What Is the *Real* Factor of Safety? 115
7.8 Bearing Capacity in 3D 116
7.9 Eccentric Loading 117
Foundations for Retaining Walls 117
7.10 Mine Collapse 119
Shallow Mines 119
Deep Mines 119
Dangers of Vertical Mineshafts 119
Longwall Mining 119
7.11 A Natural History of Caverns 120
7.12 Frost Heave and Footing Depth 120
Arctic Permafrost 121
Polygonal Ground 121
Elongated Lakes 121
Some Practical Consequences 121
Methane Release 121
7.13 When Things Go Wrong 122
Problems 123
References 123
Further Reading 123

8 The Standard Penetration Test in Foundation Engineering 125
8.1 The Empirical Approach 125
8.2 Soil Penetration Tests 125
Selective Test Depths 127
Groundwater 127
Sample Disturbance 127
The "Pocket Penetrometer" 128
Shelby Tube Samples 128
8.3 SPT in Sand 128
Depth Correction. 128
A General Depth Correction 129
8.4 Soil Mechanics of the SPT 130
What Might Be Achieved by Subtracting Blow Counts? 130
8.5 The SPT Hammers' Biggest Hits 130
Adjusting the *N* Value 130
8.6 SPT "*N*" Values and Settlement of Foundations on Sand 132
A Shallow Depth Correction 132
8.7 Pressure Bulb Correction 133
8.8 Bearing Capacity of Sand Based on an Estimated Friction Angle 135
8.9 Comparisons with Measured Settlements 135
8.10 Foundation Bearing Capacities on Clay Based on SPT or Unconfined Compressive Strength 136
Theoretical Foundation Design on Clay Based on Unconfined Compressive Strength 137
Net Bearing Pressure 137
Reducing Settlement with a Mat Foundation. 137
Summary 138
Problems 138
References 139
Further Reading 139

9 Probing with Cone Penetration Tests and the Marchetti Dilatometer 141
9.1 A Classical Approach 141
9.2 Pushing versus Driving 142
9.3 A "Friction Ratio" 142
9.4 Mechanical versus Electrical Cones 143
The Piezocone 144
Decision Time: What Are Advantages/Disadvantages of Cone and SPT?. 145
Advantages and Disadvantages of Cone Tests. 145
Piezocone and Groundwater Table 145
9.5 Fracking (Hydraulic Fracturing) 145
9.6 Example of Cone Test Data 146

9.7 Normalizing Cone Test Data for Test Depth 147
Dealing with Dimensions 147
9.8 Cone Test Data and Settlement of Foundations on Sand 148
9.9 Cone Tests and Foundations on Saturated, Compressible Clay 148
9.10 Precaution with Empirical Relationships 149
9.11 Time-outs for Pore Pressure Dissipation 149
9.12 Supplemental Cone Test Data 149
9.13 The Marchetti Dilatometer 150
Preparation for Testing 151
Soil Identifications 152
9.14 Predicting Settlement 152
9.15 A Key Question: How Can Lateral Yielding Predict Vertical Settlement? 153
Aging 153
A Dilatometer Shift in Direction of the Major Principal Stress 154
Problems 154
References 155
Further Reading 155

10 Focus on Lateral Stress 157
10.1 Lower Cost, More Convenient 157
10.2 The Pressuremeter 157
Soil Disturbance from Drilling 157
Self-Boring Pressuremeters 158
10.3 Interpretation of Pressuremeter Test Data 159
Lateral In Situ Stress 159
The Limit Pressure in Foundation Engineering 159
A Theoretical Approach 160
Use in Design 161
Soil Identifications 161
10.4 The K_o Stepped Blade 161
The Two-Chambered Pressure Cell 162
Test Sequence 163
Interpretation 163
Example 163
10.5 Summary 164
Problems 165
References 165

11 Design of Deep Foundations 167
11.1 Transferring a Foundation Load Deep to Reduce Settlement 167
11.2 When Pile Foundations Became a Matter of Necessity 167
11.3 Soils and City Planning 167
Cities and Rivers 168
11.4 Lowering of Sea Level 168
11.5 End Bearing 169
11.6 Pile Driving 169
Wood Piles 169

	The Science of Hammering	169
	Hard Driving and Brooming of Wood Piles	170
	No Lunch Breaks!	170
11.7	Tension Breaks in Concrete Piles Caused by Pile Driving?	170
	Piles Doing a U-turn	170
11.8	The Engineering News Formula	170
11.9	Pile Bearing Capacities and Load Tests	171
	Strength Gains and Slow Loading	171
	Anchor Requirements	171
	Conduct of a Test	172
	Criteria for Failure	172
	Marginal Designs	172
11.10	Analyzing Hammer Blows	173
	A Wave Equation for Driven Piles	173
	A Pile Driving Analyzer (PDA)	173
	Measuring Setup with PDA and Restrike	174
11.11	Citizen Complaints	175
11.12	Pile Load Capacities: End Bearing	175
	End Bearing on Rock	175
	Rock Quality	176
	Rock Sockets	176
	End Bearing on Sand	176
	A Critical Depth for End Bearing	178
11.13	Skin Friction and Adhesion	178
	Depth and Differential Movement	178
	Negative Skin Friction (Adhesion)	179
	End Bearing and Skin "Friction"	179
	Uplift from Expansive Clay	179
11.14	Drilled Shaft Foundations	180
	A Bad Scene	180
	Slow Demise of the Belled Caisson	180
11.15	Saving Time and Money on Load Tests with the Osterberg Cell	180
	Representative Test Results	180
	Comparisons with Top-Down Load Tests	181
11.16	Franki Piles	182
11.17	Augercast Piles	182
	Jet-Grouted Micropiles	184
11.18	Common Piles Materials	184
	Definitions of a Factor of Safety	184
11.19	Preliminary Estimates for Deep Foundation Bearing Capacity	184
11.20	Pile Group Action	188
	Pile Separation Distances	189
	Pile Group Action Formulas	189
	Batter Piles	190

Questions ... 190
References ... 191
Further Reading ... 191

12 Ground Improvement ... 193
12.1 What Is Ground Improvement? ... 193
12.2 Preloading ... 193
Enhancing and Monitoring the Rate of Settlement ... 193
A Complex System ... 194
12.3 Compaction ... 194
Vibratory Compaction ... 194
Deep Dynamic Compaction (DDC) ... 194
Blasting ... 195
Side Effects from Compaction ... 195
12.4 Soil Replacement or Improvement ... 195
Stone Columns, Aggregate, and Mixed-in-Place Piers ... 195
12.5 Grout Materials ... 197
12.6 Grout "Take" ... 197
12.7 Rammed Aggregate Piers ... 197
A "Saw-Tooth" Stress Pattern ... 199
Temporary Liquefaction ... 199
Tension Cracks Outside the Liquefied Zone ... 199
12.8 A Hypothesis of Friction Reversal ... 200
Conditioning ... 201
Friction Reversal and Overconsolidation ... 201
12.9 Advanced Course: Application of Mohr's Theory ... 201
Lateral Stress and Settlement ... 202
Is Excavation Permitted Close to RAPS? ... 203
12.10 Further Developments ... 203
RAPS as Anchor Piers ... 203
When Soil Does Not Hold an Open Boring ... 203
Low-Slump Concrete Piers ... 203
Sand Piers ... 203
Questions ... 203
Reference ... 204

Appendix: The Engineering Report and Legal Issues ... 205

Index ... 207

Preface

The thread of learning is strengthened through understanding.

Soil is the most abundant construction material, and also the most variable. Early engineering tests of soils involved the resistance to jabbing with a heel or probing with a stick. Probing then developed along two different approaches, hammering and pushing. Both can provide useful information, but the tests do not accurately simulate soil behavior under or near a foundation.

Targeted Tests

A targeted test is one that is directly applicable for design. An example is a pile load test that relates settlement to the applied load. A load test also can be continued to determine an ultimate bearing capacity. A *plate bearing test* can similarly model a shallow foundation, but scaling down makes the results less directly applicable.

A third approach is to obtain and preserve soil samples in their natural state and test them in a laboratory. The problem then becomes how to collect a soil sample without disturbing it, as even the removal of a confining pressure can effect a change.

An Early Targeted Test

The laboratory consolidation test devised by Karl Terzaghi was targeted to measure soil behavior as it may influence foundation settlement. Observations and measurements made during the tests then led to an important spinoff, the concept that pore water pressure subtracts from normal stress and therefore from friction. That now is considered by many to be the entry point for modern soil mechanics.

A Simple Targeted Test

The plastic limit test must be one of the simplest soil tests ever devised, and results are part of most engineering soil classifications. The test uses hand power to roll out, bunch up, and re-roll threads of soil until it dries out and crumbles. The transition moisture content is the plastic limit. It not only depends on a soil clay content but also on its clay mineralogy, and the test was devised long before it became recognized that there is a clay mineralogy.

Two Requirements in Foundation Design

Requirements are as follows: (1) Settlement must be uniform and must not be excessive, and (2) a foundation must not punch down into the ground in a *bearing capacity* failure. If a near-surface soil is not adequate, deep foundations can transfer loads downward to bear on rock or in more competent soil. A complication for deep foundations is that they can derive support from two sources, end-bearing and side friction, and the two contributions are not separated with ordinary top-load tests. They can be isolated by using an expandable *Osterberg cell* to push up from the bottom. Pile behavior and integrity also can be examined with impacts and sound waves.

A New Role for Lateral Soil Pressure

Laboratory triaxial shear tests define relationships between lateral confining pressure and soil strength and bearing capacity. Field tests have led to the discovery that a high lateral pressure imposed on saturated soil can work a temporary change in the soil behavior, and the change can be an important factor affecting foundation settlement. That development is given special attention in the last chapter of this book.

Soil Origins and Clay Mineralogy

One mistake is one too many, but mistakes happen. In foundation engineering a mistake sometimes can be attributed to a disconnect between engineering purpose and site geology. Most soil is hidden away, and geology and soil science, which emphasizes changes caused by weathering, can reveal where and what to look for. For example, expansive clays that cause no end of engineering problems are far more common than can be shown on small-scale engineering soil maps. The geotechnical engineer who is not cognizant of geological relationships and engineering consequences is riding on one wheel.

The Engineer as Teacher

Case history. An architect designed a building with exterior walls of Italian marble, and was in no mood to spend money for deep foundations or anything else that "would not show." He had to be convinced that without deep foundations, the consequences would show.

Introduction

Some Heroes in Geotechnical/Foundation Engineering

Archimedes (287–212 BC) famously discovered "Archimedes Principle" of buoyancy, which affects soil weight and frictional resistance to sliding. He was killed by a Roman soldier who had no appreciation.

Charles-Augustin de Coulomb (1736–1806) was a French military engineer, and while being in charge of building a fort on the island of Martinique he observed that sand grains must have friction or they would not make a respectable pile. He also reasoned that clay must have cohesion or it would not stand unsupported in a steep bank. Those observations led to the "Coulomb equation" for soil shear strength. Over 100 years later, Karl Terzaghi added the influence from pore water pressure that tends to push grains apart.

Coulomb also derived an equation for the lateral force from soil pushing against a retaining wall. The equation, and a later equation proposed by Rankine, puts the maximum soil pressure at the base of a wall but tests conducted by Terzaghi indicate that it is more likely to be zero. That is no small error because raising the height of the center of pressure increases the overturning moment, which makes the Coulomb and Rankine solutions the unsafe side.

Coulomb's Law

After retiring from the Army, Coulomb entered a contest to invent a better marine compass. He did not win the contest but invented the torsion balance that substitutes twisting of fine wires for knife edges. Coulomb then experimented with his instrument to measure tiny forces from electrical charges, electricity being big at the time, and discovered that forces between two electrically charged particles depend on *square* of the separation distance. Coulomb's Law also governs space travel and orbiting distances of satellites.

William John Macquorn Rankine (1820–1872) was a professor at the University of Glasgow. He was most famous for his analysis of the thermodynamics of steam engines, but he also had a simple solution for soil pressures against retaining walls. He defined an *active state* for soil that is acting to retain itself, and a *passive state* for soil that is being pushed. Rankine's and Coulomb's analyses can give the same answers, but both have a limitation.

Christian Otto Mohr (1835–1918) was a German bridge engineer and a professor of mechanics at Stuttgart and Dresden. He devised the "Mohr circle" graphical method for depicting soil stresses, and the "Mohr envelope" defines stress conditions for shear failure. It supports Coulomb's soil shear strength equation.

Ludwig Prandtl (1875–1953) was a professor at the University of Hanover, most famous for his contributions to aerodynamics. He also developed a theory for the resistance of metal to penetration by a punch based on a curved failure surface called a *log spiral*.

Karl Terzaghi (1883–1963) was from Austria and was educated in mechanical engineering. However, he also was interested in geology and became a professional geologist. He then used an engineering approach for soil problems, for example, by applying Prandtl's *log spiral* to shallow foundation bearing capacity, a theory and approach that still are widely used. As a professor at Robert College in Turkey, Terzaghi devised the consolidation test and theory for predicting foundation settlement. Those observations led to defining soil shear strength in terms of *effective stress* that takes into account the influence from excess pore water pressure.

Terzaghi also observed that because clay particles must be soft and yielding, contact areas between particles can be expected to vary depending on the contact pressure, which might explain the linear relationship between friction and normal stress. It is the concept that made its way back into mechanical engineering to explain friction. It also can explain the function of a lubricant, to keep surfaces separated.

Geotechnical engineering has grown and continues to grow, and many investigators and practitioners continue to make important contributions. Broad interests, curiosity, imagination, and an interest in working with a complex and somewhat unpredictable natural material are part of the toolkit.

Further Reading

Bowden, F. P., and Tabor, D., *The Friction and Lubrication of Solids*, Oxford University Press, Oxford, UK, 1950.

Casagrande, A., "Karl Terzaghi—His Life and Achievements," In *From Theory to Practice in Soil Mechanics*, L. Bjerrum, A. Casagrande, R. B. Peck, and A. W. Skempton, eds. John Wiley & Sons, New York, 1960.

Handy, R. L., "The Arch in Soil Arching," *ASCE Journal of the Geotechnical Engineering Division*, 111(GT3):302–318, 1985.

Terzaghi, K., *Theoretical Soil Mechanics*, John Wiley and Sons, Inc., New York, 1943.

Karl Terzaghi (1883–1963). Pencil sketch by Tauseef Choudry.

CHAPTER 1

Defining What Is There

Geology and Foundation Engineering

1.1. The Three Most Common Construction Materials

Concrete has a recipe, steel is made to order and goes by number, and soil and rock are what is there. A first requirement in foundation engineering therefore is to determine and characterize what is there. That requires knowledge or at least familiarity with site geology.

For example, soils of river floodplains are likely to occur as sedimentary layers. A common sequence is clay layers on top of sand layers on top of gravel, as flow velocities decreased during stages of deposition. Soils deposited by winds are more likely to transition horizontally, from sand dunes adjacent to a source to thick, highly erodible deposits of silt that has such an open structure that when saturated with water can collapse under its own weight. With increasing distance from a source the silt is transitional to clay that is particularly troublesome because it is expansive and can lift building foundations in the presence of water.

Procedures used for identifying, boring, probing, sampling, and/or testing vary with different kinds of deposits because of the variability and focus on particular engineering properties. Core samples obtained by pushing a steel tube into the soil are commonly called "undisturbed," but the term is shielded by optimism. A soil that is relieved of existing pressure will respond by simply expanding, so it, to some degree, is disturbed. It also is not possible to accurately reproduce field conditions in a laboratory if, as often is the case, those conditions are not known and are difficult to measure. Many important engineering soil properties are inherited, for example, from having been buried under a thousand meters of glacial ice or a hundred meters or more of soil that has been removed by erosion.

Soils usually are investigated with borings, but there can be no guarantee of what engineering perils may exist between the borings. This limitation is included in every geotechnical report, and usually is written with the assistance of an attorney. Geological awareness can help to make sense out of a situation and can be critical.

Supplementary data can be obtained with geophysical seismic (ground echo) or electrical resistivity tests, and with ground-penetrating radar. Simplest to interpret is the radar that prints out a running log as the instrument is being pulled over the ground, but a limitation is that the depth scale depends on the moisture content, and penetration is limited.

Most important can be observations of soils and rocks exposed by erosion and occurring in outcrops and excavations. Airphotos and drone photos can reveal patterns that are easily overlooked from the ground. Interpretation depends on a sound knowledge and appreciation for geological origins.

1.2. Two Classes of Foundations

Foundations are described as shallow if they bear on near-surface soils or rocks, and deep if they extend down to firmer soil layers or rock. Deep foundations are more likely to be used to support heavy structures such as multistory buildings, and can be effective even if bedrock support is not available. Shallow foundations may be suitable for supporting lighter weight structures, depending on the firmness of the soil.

Support of Deep Foundations

Two sources of support for a deep foundation are end bearing at the base and shearing resistance along the sides, usually referred to as a *skin friction.* As the contributions involve different soil properties and are unlikely to peak out together, they are analyzed and/or measured separately, as side resistance often peaks out and starts to decline before end bearing is fully mobilized. A further complication is that if the ground settles, usually as a consequence of lowering the groundwater table, then skin friction is reversed, so it pushes down instead of up.

Expansive Clays Can Be Expensive Clays

Near-surface soils in many areas of the world often include clay minerals that expand when wet and shrink when dry, affecting pavements and foundations. The problem is intensified because dry weather is preferred for construction, when the clays are dry and deceptively hard but poised and ready to expand. Floors and foundations usually are raised unevenly, so walls develop diagonal cracks, and door and window frames can be distorted so they no longer are rectangular.

Expansive clays are the most costly problem in geotechnical, highway and foundation engineering, with a tally running into billions of dollars annually in the United States alone. But there are remedies and solutions.

End Bearing on Rock

Solid rock can be an ideal support for foundations but basement excavations may be too costly to be practical. Solid rock can lay buried underneath weathered rock and rock fragments and/or geologically younger soil deposits, so these can be penetrated with borings or driven piles that may require end protection with hardened steel tips.

A particularly serious problem can be shallow underground caverns or mine openings that remain undetected until a heavy load is applied. Caverns are created in limestone where infiltrating seepage water that has been rendered slightly acidic by dissolved carbon dioxide and concentrated at a groundwater table. The caverns therefore may be relatively deep and difficult to detect. In geological time as nearby valleys are eroded deeper, the groundwater level is lowered, so caverns become accessible for spelunking. However, vertical channels appropriately called "glory holes" may connect different cavern levels.

Even shallow limestone may hold some surprises if it has been weathered along vertical fractures that become filled with clay. As this is associated with surface weathering and development of a "soil profile," it is discussed later in this chapter. It is the shallow clay pockets, caverns, and mine openings that are most likely to cause problems.

Ground Improvement

Ground improvement means to improve what is there. A simple procedure is to let a pile of soil stand in for a future foundation until settlement stops, then remove the soil and build the structure. This procedure is particularly useful when a series of similar structures such as apartment houses are to be constructed so that after settlement is complete, the soil can be moved on to the next site. Engineering still is required to determine an appropriate preload pressure, measure settlement, and determine if a well system may be required to assist in the removal of water as it is squeezed out of the soil.

Dynamic compaction: Layers of soil can be spread and compacted with rollers or vibrators to create a structural fill. This procedure can dominate cut-and-fill operations for roads and highways, and can be used for foundations. Careful selection of a satisfactory fill soil is required, and standardized test procedures are used to determine appropriate soil moisture content and acceptance criteria for testing and compaction.

Deep dynamic compaction is more likely to be used to process soil in situ in preparation for a future foundation load. It involves using a crane to repeatedly lift and drop a heavy weight to pound the soil into submission. It is best adapted to rural areas.

Chemical soil stabilization can be achieved by mixing soil, Portland cement, or chemical lime prior with soil and compacting it in layers. For deep in situ stabilization, the lime can be introduced into open borings or mixed with the soil in situ in the borings. Lime reacts chemically with expansive clay minerals, so they harden and become non-expansive.

Procedures for ground improvement have received considerable attention in recent years, and are discussed in more detail in the last chapter in this book.

1.3. Residual Soils

Granite mountains are the ultimate source for most sand. Most granite is igneous rock, which means that at one time it was molten, and then slowly solidified at great depth. Therefore, individual crystals are sand-size and larger. Clear grains are quartz, and pink or white grains are feldspars that are more readily weathered to form clay. As feldspars chemically weather to clay, the grains expand, and granite becomes separated into grains of sand.

Travel Is Wearing

Sand is readily moved by gravity, wind, or water. Close to the source, the sand usually has about the same color as granite because it has the same mineralogical composition, about 25 percent quartz and the remainder feldspars. Farther from a

FIGURE 1.1 Granite is a major component in most mountain ranges and disintegrates along cracks to leave rounded boulders and sand. (Image source: *Geotechnical Engineering: Soil and Foundations Principles and Practice* by Richard L. Handy and Merlin G. Spangler. McGraw-Hill Educations © 2007.)

source, as feldspars weather and are degraded the mineral percentages are reversed, about 75 percent feldspar and the remainder quartz, with a characteristic tan color from grains being coated with iron oxides and clay. An exception is "Ottawa sand" that is a fossil beach sand and is almost pure quartz. One use is to make glass.

Weathering along cracks in granite leaves rounded boulders, as shown in Fig. 1.1. They obviously have not been rounded by rolling along in streams, as commonly assumed. As a general rule, rocks form mountains, which weather and disintegrate into soil that is moved downhill by water and gravity into adjacent valleys where they can be further modified by weathering or moved along by wind and water.

1.4. Soil Layers Created by Weathering

Topsoil "A Horizon"

Topsoil is preferred for gardening but not for engineering, as it contains organic matter that can separate grains and weaken the soil. Topsoil typically is 1–2 ft (0.2–0.5 m) thick unless eroded. At a construction site, topsoil usually is stripped off and saved for later use as a top dressing for lawns. Topsoil may be black from organic matter, and brown or red-brown from iron oxide coatings on soil grains. When developed under trees it can have a thin gray or white layer because of intense weathering from acid soil conditions.

Figure 1.2 A weathered soil profile in expansive clay: Dark, organic *A horizon* topsoil about 8 in. (20 cm) thick on top of brown, clayey *B horizon* subsoil that has a blocky structure indicative of expansive clay. *Olton soil series in western Texas.* (Image source: USDA.)

Subsoil "B Horizon"

Clay that is created by weathering in the overlying A horizon can be carried down with infiltrating rain water to be deposited and concentrated in a relatively thicker subsoil layer called a "B horizon." In temperate climates, B horizons commonly contain concentrated *expansive clay minerals* that shrink when dry and expand when wet. This can cause major problems in engineering.

Shrinkage Cracks and Blocky Structure in Expansive Clays

Vertical shrinkage cracks can define an "active layer" of shrink–swell cycling in expansive clay soils. In a B horizon the cracks can intersect to form a characteristic "subangular blocky" soil structure, as shown in Fig. 1.2. Blocks are coated with thin layers of expansive clay called "clay skins" that prevent bonding, so the soil is avoided for use in foundation engineering.

Case history. Expansive *B horizon* clay soil was recognized and removed from a building site. The pile of soil was not recognized as being expansive and was used as fill soil for another building site, with predictable consequences.

1.5. Vertical Mixing in Expansive Clay

In areas with extended dry seasons, vertical shrinkage cracks can extend a meter or more deep and can define an *active layer* in expansive clay. Cracks are an open invitation for debris and soil that slough off and prevent closing. Repeated open–close cycling then can build up sufficient lateral stress that exceeds the soil unconfined compressive strength, and the soil shears *along inclined planes.*

Shrink–swell cycling, therefore, can eventually mix the A–B horizons into a single thick layer that is expansive and black. The soils have been appropriately referred to as *black cotton soils.* The scientific name is *Vertisol,* for vertical mixing. The soils are bad news for engineering. A clue can be above-ground burials in cemeteries. Various methods can be used to deal with such soils and are discussed in Chap. 4.

1.6. Influence from a Groundwater Table (or Tables)

The level to which water rises in a well defines the groundwater table. It is replenished by seepage so the groundwater level tends to be a weakened expression of hillside surface elevations. Saturation of soil under a groundwater table reduces soil unit weight about one-half; therefore, it can have a major influence on engineering uses as well as contributing to wet basements.

The elevation of a groundwater table obviously is important in engineering, and can be measured from the water level in borings that have been left open for a day or more. The measurements usually are made with a tape that employs an electrical contact.

Groundwater Table and Soil Color

The elevation of a groundwater table can change seasonally depending on rains. Soil below a permanent groundwater level develops a diagnostic gray soil color and is referred to as "unoxidized," as the gray color is attributed to a lack of oxygen dissolved in the water. Infiltrating rainwater contains dissolved oxygen that can react with iron compounds that stain soil grains to a shade of tan or brown. A seasonally changing groundwater level creates a mottled mixture of gray and brown, sometimes with vertical lines of rust concentrated along former root channels.

A color determination has obvious relevance in engineering as it can indicate seasonal variations in the level of a groundwater table. The examination of soil color should proceed and be recorded soon after the soil has been removed from a boring because it can rapidly change upon exposure to air.

Some guidelines for soil color are listed in Table 1.1. *It will be noted that soil color is not revealed by probing.* A more detailed identification can be made using color charts in the Munsell system, and charts showing only colors commonly found in rocks and soils are available from suppliers.

A Perched Groundwater Table

Downward seepage of water through soil may be impeded by a buried layer of clay to create a "perched" groundwater table that is separated from a deeper and more permanent groundwater level. The clay layer often will represent a former ground surface

Black, dark brown	Organic topsoil or *A horizon*. Avoided in engineering.
Thin, light gray or white	Indicates acidic conditions in soil developed under forest.
Tan or brown	Most common, having been oxidized by exposure to air and therefore above a groundwater table.
Mottled brown and gray	*Fluctuating groundwater table*. Important to recognize in engineering, seasonal changes in buoyant support reduce shearing resistance.
Gray	(a) Most commonly indicates reducing conditions from a lack of oxygen below a permanent groundwater table. (b) May be "fossil" in geologically young glacial soils that have been highly compressed and rendered impermeable by a heavy weight of glacial ice.
Blue or green	Excessive reducing conditions indicating marshy conditions. Can be an important clue to a gas leak. Also can occur in soil after prolonged contact with a bituminous pavement.
White, crusty	*Caliche*: Concentrations of calcium carbonate formed in near-surface soil where the rate of evaporation exceeds the rate of precipitation. Characteristic of near-surface soils in an arid or semiarid climate.

Table 1.1 Some Guidelines for Soil Color

with buried *A* and *B* horizons that are called *paleosols,* for ancient soils. A perched groundwater table can be troublesome, as it can drain into an open excavation.

1.7. Intermittent Recycling

Many soils used in engineering are *sediments,* with properties that are defined by their geological origins. In geological time, sediments become compressed and cemented to form sedimentary rocks. Most common is *shale,* which typically is gray, dense, and thinly layered from having been deeply buried prior to being exposed by geological erosion.

Most shales are deposits from shallow seas that covered areas of continents during past geological time. Rocks that are not thinly layered and are composed of clay are *claystones.* Shale usually is dominant, and often is interlayered with sandstone, limestone, and coal.

Shales of intermediate geological age are less likely to be thinly layered and may contain expansive clay minerals and occasional dinosaur tracks. Thin layering is not a criterion for expansive or non-expansive clay.

1.8. Soil Types and Foundations

The simplest foundations are *slab-on-grade,* concrete slabs that are flat and level. If a foundation slab covers expansive clay, the slab will restrict evaporation, and therefore moisture accumulating under a central area will expand the clay and lift the center part of a structure more than the edges. Expansive clay problems are discussed in Chap. 4. *Shallow foundations* extend down through topsoil, but still can be affected by expansive clays.

Column foundations usually are square but can be round. As discussed later in this chapter they can be hit-or-miss when founded on weathered limestone. *Wall foundations* are linear and more likely to bridge across weak areas. Shallow foundations are commonly used for supporting lightly loaded structures.

Deep foundations initially were straight tree trunks that were stripped of bark and branches. They usually are driven upside-down to take advantage of a natural taper, and still are widely used. Driven piles used to support heavy structures are more likely to be steel or concrete. Steel piles can be pipes, H-beams, or hollow and tapered.

Concrete piles can be driven, or they can be larger in diameter and bored-and-poured. Resistance to lateral forces can be increased by incorporating a "cage" of steel reinforcing that is lowered into the concrete before it sets.

In caving soils that do not hold an open boring, an *augercast pile* is created by twisting a hollow auger that is the full length of a pile into the ground, so soil between the spirals holds the boring open. As the auger is raised, cement grout is pumped down through the center pipe to create a pile from the bottom up. Positive fluid pressure is maintained to prevent caving.

Two classes of deep foundations are end-bearing, which transfer load down to a hard stratum such as bedrock, and *friction* that transfer load to soil that is in contact all along the surface. However, the definition is not exclusive because both mechanisms can contribute, but is highly unlikely that the two resistances will develop and peak out together.

Three consolidation classes of soils: As soil tends to consolidate under its own weight, it typically becomes more supportive with increasing depth. A soil that has been consolidated to equilibrium under existing overburden pressures is said to be *normally consolidated.* Its density and unit weight, therefore, increase with depth.

A soil that has been consolidated under a prior larger overburden pressure is *overconsolidated.* Overconsolidation is advantageous because it can reduce and even prevent significant foundation settlement if the foundation pressure is less than the prior overburden pressure. Some *quasi-elastic* settlement will occur. Overconsolidation can occur with a single emergence–submergence cycle of a groundwater table, so the ideal, normally consolidated soil may be difficult to find in nature.

A soil that is *not* in equilibrium with the existing overburden pressure is said to be *underconsolidated.* This obviously is a potentially unstable condition because if conditions change, the soil may consolidate. In recently deposited soil, the time after deposition may not be sufficient to allow drainage of excess pore water. In that case consolidation is on-going and can be expected to speed up with additional loading. The other common cause for underconsolidation is most likely to be encountered in wind-deposited loess soil, where grains are pulled together by negative (capillary) pressure that can be lost upon saturation with water.

Influence of a Groundwater Table

Even though water can only occupy pore spaces between soil grains, about half of the weight of soil that is under water is supported by buoyancy. As the elevation of a groundwater table depends on the availability of water, it can vary seasonally. However, it is a first-time lowering of a groundwater table that can be most damaging because the soil may be subjected to a load that it has not experienced before. This is most likely to occur in cities, where a surface cover drains carry water away instead of allowing it to penetrate into the ground.

As lowering of a groundwater table removes buoyant support for the soil, even deep foundations such as piles can be affected because if the soil settles more than the pile, it can create downdrag or "negative skin friction" that can overload the pile. Even if a deep foundation is designed to accommodate negative skin friction, settlement of adjacent soil can affect sidewalks, sewers, and utility lines.

Pull-up of Deep Foundations by Expansive Clay

Seasonal cycling of expansive clay in contact with a deep foundation can incrementally jack it up out of the ground. The problem is particularly relevant for bridges, as they are designed to carry heavy truck loads and remain without the extra load most of the time. Pullout can be prevented by coating the part of the foundation element with a soft, semiliquid layer such as bitumen. Another method is a "bell-bottomed" steel-reinforced concrete pier that acts as an anchor.

1.9. Agricultural Soil Maps

County soil maps in the United States prepared and published by the USDA are mainly used for agriculture, but these are also a valuable resource for geotechnical engineers because investigations are made on-site and soil boundaries are mapped with the assistance of air photos. The reports now are prepared with the participation of engineering agencies such as state DOTs, and engineering soil classifications and data are included in the reports. Unfortunately the reports usually stop at city boundaries, but they still can be used for expanding suburb areas.

The Soil Series

Formal coursework in soil science or pedology is helpful for geotechnical engineers but is not available at some universities. The basic mapping unit is the *soil series*. Definitions and boundaries are closely defined. A soil series designation depends in part on drainage, so several series can exist on the same hillside.

Series names can identify and distinguish important engineering properties including expansive and non-expansive clays, river sand that is dense and dune sand that is loose, and soils that are subjected to flooding. Soil series are given local names and code letters on maps, as shown in Fig. 1.3. Maps are prepared by soil scientists who can read the topography, walk the fields, and probe and describe the soils. Print copies of the maps are available from local offices of the USDA, and also are available online from the U.S. Government Printing Office.

1.10. Distinguishing between Alluvial Soils

Figure 1.3 shows an example of a *meandering river,* a river that automatically adjusts its length in order to achieve a stable downhill gradient. Banks around the outer sides of curves are undercut by erosion so they cave off in near-vertical failures. River approaches to bridges often are protected with sheet piling or by dumping stones or broken concrete.

A *braided river*, as shown in Fig. 1.4, transports more sand and gravel than it can carry, so channels become plugged and divided to create a braided appearance. The downhill gradient remains high so flow is rapid and the potential for bank erosion is high.

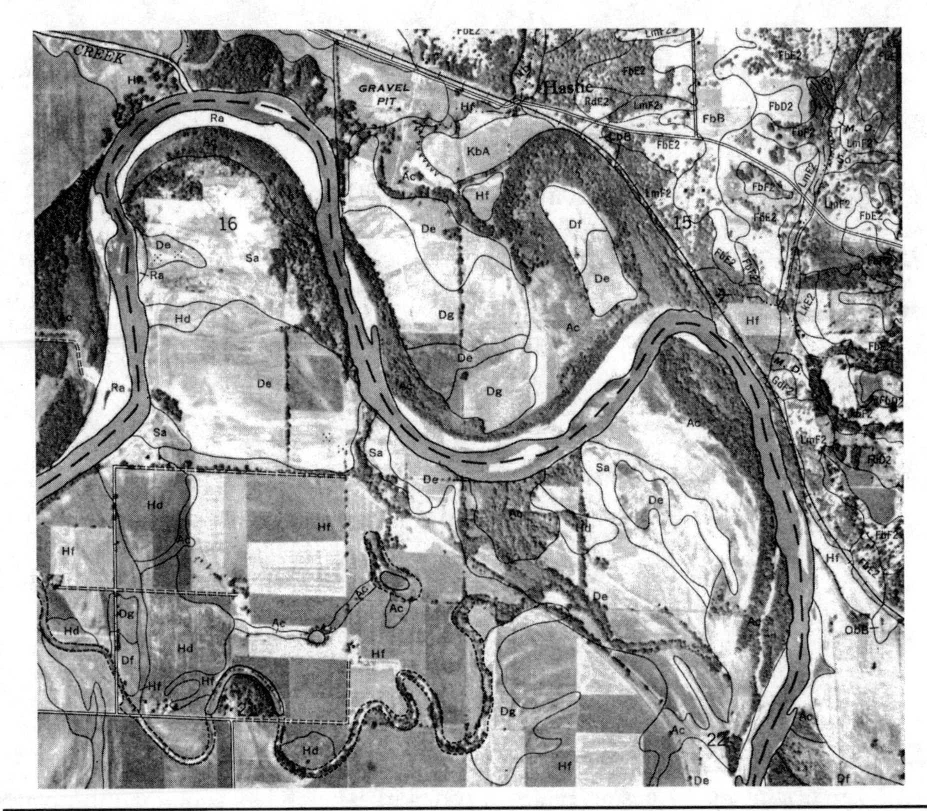

Figure 1.3 Agricultural soil map superimposed on an air photo. River meanders enclose light-colored sand *point bars*. At the right-center, the river has taken a shortcut and left a *meander cutoff and oxbow lake*. When filled with clay, oxbow lakes are some of the least desirable soils commonly encountered in engineering. (Image source: U.S. Dept. of Agriculture Soil Survey Series 1953, No. 9, Polk County, Iowa, Sheet No. 15.)

Rivers and Continental Glaciation

Downhill gradients of continental glacier ice surfaces were sufficient to cause them to creep across half a continent, until the rate of melting at glacial margins equaled and eventually exceeded the rate of ice advance. They trapped and held so much water that sea level was lowered about 400 ft (120 m), indicated by canyons that were eroded where rivers entered the sea and were filled with sand and gravel glacial outwash as sea level came back up. Foundations for bridges crossing major rivers such as the Mississippi, therefore, are founded in glacially derived sand and gravel. Drowning of rivers that were not carrying glacial outwash created fiords, or when filled with fine-grained sediments became estuaries.

Meanders and Cutoffs

Sand "point bars" occupy the space enclosed by meander loops, and as shown in Fig. 1.3, the most recently deposited sand usually is exposed along inner edges of the meander loops. During periods of high water, the sand can be covered by a thin layer of clay. Sometimes a small channel will take a shortcut across a point bar and create its

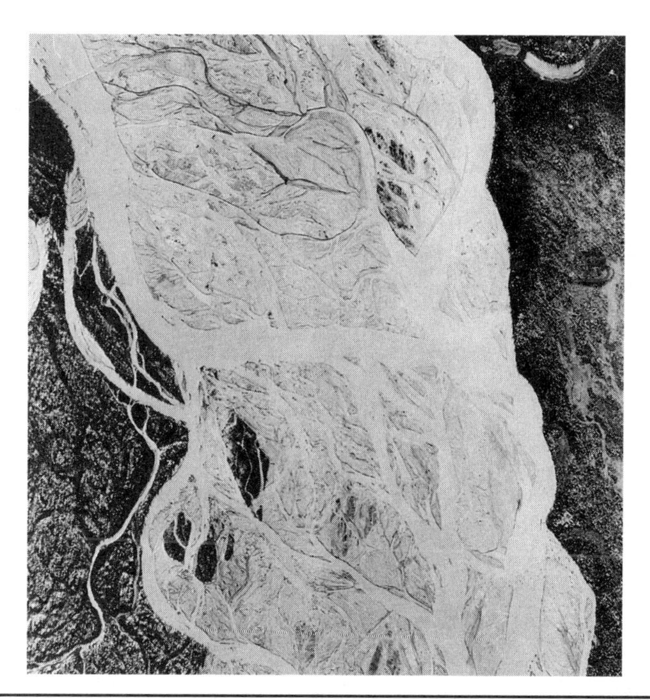

FIGURE 1.4 Air photo of a braided river carrying sand and gravel outwash from a melting glacier in Alaska. The area at the right is a terrace that was created when the river cut deeper, and shows similar scroll patterns. (Image source: *Geotechnical Engineering: Soil and Foundations Principles and Practice* by Richard L. Handy and Merlin G. Spangler. McGraw-Hill Educations © 2007.)

own scaled-down meanders, as shown in the lower part of Fig. 1.3. The shortcuts sometimes became traps for steamboats if the river level unexpectedly went down.

After periods of flooding, saturated soil lining outer banks of meanders tends to cave off into the river, causing meanders to snake their way downstream. As the outer bank moves, the inner bank follows by depositing sand point bar. When a meander encounters a more resistant stratum such as an older, clay-filled oxbow, its migration may be slowed sufficiently that the following meander overtakes it to create a *meander cutoff.* When plugged at the ends it becomes an *oxbow lake.*

Oxbow Lake Clay

An oxbow lake remains at about the same elevation as the nearby river. It readily becomes filled during periods of high water and traps even fine-grained clay when the river level goes down. Sedimentation rates are reduced by buoyancy and a low specific gravity and fine particle size of expanded clay, so it accumulates as a soft, semi-liquid mass. After repeated episodes of flooding it can fill the lake to create some of the weakest, most compressible soils commonly encountered in foundation engineering. During the heat of a summer, the exposed surface may become desiccated into a hard crust with a pattern of intersecting shrinkage cracks.

The bottom contours and hence the thickness of an oxbow lake clay is that of the river channel when the water level was highest and channel erosion was deepest, at the time of the cutoff. Compounding the problem is because after an oxbow lake is completely filled with sediment it may be buried under more recent deposits, so it is not possible to detect and confirm it without exploration drilling and/or geophysical methods.

Case history. The approach embankment for a highway overpass extended out onto an oxbow lake clay bordering a major river valley. The oxbow was obscured by later deposits and partly covered by a road embankment, but nevertheless was identified by the agricultural soil series on a published USDA county soil map. The exploration drilling program was limited to a single boring that missed the oxbow. The nose of the embankment caused a bearing capacity failure that pushed against and bent H-beam piles driven to support a highway overpass.

Alluvial Fans

Oxbow lakes that occur along river floodplain margins are likely to be covered with coalescing *alluvial fans*, fan-shaped deposits of sand and silt caused by a change in gradient of streams emerging from nearby uplands. The fans slope down to merge with the floodplain, and, being at a higher elevation than the floodplain, are commonly used to support roads and other structures to reduce the frequency of flooding. However, channels of the streams swollen with torrential rains can be aggressively eroding and take out sections of a fan in minutes.

Natural Levees

Natural levees form along outer rims of meander loops when a river spills over its banks during floods. Sand and silt are deposited so the levee surface is higher than that of the floodplain. Therefore, even though they occasionally can flood, the levees are preferred building sites. The oldest part of New Orleans is on a natural levee. Flooding can be reduced or prevented if a river is brought under control with dams and reservoirs.

Slack-Water (Backswamp) Floodplain Deposits

An overbank natural levee deposit is transitional to "slack water" or "backswamp" clay deposits that tend to dominate a meandering river floodplain to cover and obscure older point bar and oxbow lake deposits. Rivers can carry and deposit expansive clay on floodplains and in rice paddies in tropical areas where the upland soil is highly weathered, red, and non-expansive.

Air Photo Interpretation

Figures 1.3 and 1.4 illustrate the usefulness of air photo interpretation to identify meandering and braided rivers and their deposits. In Fig. 1.3, meanders and their enclosed point bar sand deposits that are well drained and appear lighter, and at the upper right is a meander that has been cut off and is in the process of becoming a clay-filled oxbow lake. A railroad is running along on top of a natural levee. The meander size and channel size are interrelated, and "chute," a small channel cutting across a point bar, has its own meander pattern. A gravel pit identified near the top of the photo is likely to be a terrace consisting of glacial outwash, so other similar terrace remnants probably exist and can be sources for aggregate. As gravel terraces are deposited during glacial melting when the river was braided, they should have a steeper downstream gradient so that they merge with and are covered by the more modern floodplain.

In Fig. 1.4, the braided pattern is characteristic of actively deposited sand and gravel, and a similar braided pattern can be observed that can identify older deposits in the adjacent tree-covered terraces.

1.11. Wind-Deposited Soils

Sand Dunes

Sand grains that are carried by wind skip along across a ground surface until they encounter vegetation or other obstacle to pile into a sand dune. Grains that hop up the windward side drop over the top edge to slide down the lee side at an angle appropriately called the *angle of repose.* The angle is approximately the same as the friction angle for quartz, about 25°. Pounding by skipping sand particles can create a shallow crust that makes the rounded windward side more suitable for dune buggies than more steeply slopping loose sand on the lee side.

Sand that is carried around lateral ends of a dune can create "tails" that give a dune a characteristic shape and feed into other dunes, developing a *dune tract.* When dunes are on the march, the best defense is to stay out of the way. Second choice is to do battle with vegetative cover at the source and which can tolerate difficult conditions. Stable dunes often are pockmarked with *blowouts* that expose bare sand.

Sources for sand: The most common sources for sand are deserts, beaches, and braided river floodplains. The latter can be associated with outwash from melting of continental glaciers. "Cliff-head" dunes occupy stable positions where winds sweep upward over areas bordering beaches and floodplains. Point bar sands on meandering river floodplains are more likely to be covered with trees and other vegetation, and are surrounded on three sides by the river.

Eolian sand is fine-grained, and grains tend to be uniform in size so it is a "poorly graded" engineering soil that does not compact well.

Where there is a continuous source of sand, a downwind march or expansion of dunes is inevitable. Roads that go where dunes are unavoidable will require the occasional attention of a road grader.

Eolian Silt Deposits

Sand grains are large and heavy enough that they do not remain in suspension. They bounce over a ground surface, whereas silt grains and connected clay particles are carried in suspension as dust, so they are distributed and deposited across large areas. Dust storms prevailed in the Pleistocene, and winds blowing nearly parallel to river valley sources piled up large thicknesses within a few kilometers of the valleys. A Native American name for the Missouri River valley means "valley of smoke," and radiocarbon dating indicates that dust deposition ended about 12,000 years ago.

Deposition of silt particles carried by winds can be hurried along by rains, or silt grains may only settle out and infiltrate down into a vegetative cover. As deposition continues and engulfs the vegetation, the density remains low and may preserve vertical root channels that increase permeability.

The name given to wind-deposited silt is Löss, from German for loose. It has been Anglicized into *loess,* variously pronounced as luss, *lo-ess,* and *lerse.* Thick loess deposits are *underconsolidated and collapsible* when saturated with water. At one time it was assumed that loess grains are cemented with calcium carbonate cement, but this

Figure 1.5 Author on a rope, sampling wind-deposited loess soil adjacent to its Missouri River floodplain source. Deposition during the last glacial advance began about 25,000 years ago and ended about 12,000 years ago. (Image source: *Geotechnical Engineering: Soil and Foundations Principles and Practice* by Richard L. Handy and Merlin G. Spangler. McGraw-Hill Education © 2007.)

concept is not sustained by microscopic observations. Instead the grains appear to be held together by coatings of clay and by capillary tension that pulls outward on menisci to create negative pore water pressure. Both actions will be weakened by the presence of water.

An example of thick, collapsible loess is shown in Fig. 1.5. Structures can be safely supported on collapsible loess soil but only as long as it remains well drained.

Transition to expansive clay: Loess deposits cover about 10 percent of the earth's land surface. However, loess rapidly becomes finer and more clayey with increasing distance from a source, and at longer distances it develops a weathered soil profile and transitions into expansive clay. This is discussed in more detail in Chap. 4.

1.12. Landslides

Landslides can be sudden, unexpected, dangerous, and devastating. They usually occur during wet weather when soil is saturated, and pressure from pore water against soil grain surfaces pushes them apart and reduces contact friction.

Landslides occur naturally on hillsides sculptured out by geological erosion, and can be triggered by man's activities. The most common causes are from adding weight at or near the top of a hill, removing lateral support near the bottom, or interrupting natural drainage.

Landslide Scarps

A common diagnostic feature for landslides is a near-vertical *scarp* of bare soil that marks a temporary upper boundary and is exposed slip surface. As downhill movement

FIGURE 1.6 As landslides move downhill, the area involved expands uphill to create a series of near-vertical scarps. (Image source: Author photo.)

along the scarp takes lateral restraint away from the soil, a parallel tension crack often will form that delineates the path of the next scarp as the slide area expands uphill and more soil joins the slide.

Landslide scarps therefore can form a series of steps that are slightly rotated inward so they trap and hold rainwater that seeps into the soil and further aggravates the situation. A series of scarps are shown in Fig. 1.6, where upward expansion of the slide area continued until it took a line of new houses.

Case history. As a landslide scarp can create a bad impression for potential real estate buyers, it may unethically be covered up. That usually will void a sale and/or lead to a lawsuit.

The landslide scarp in Fig. 1.6 was buried under several meters of loose fill soil that was inadequate to support the building foundations, but was not interpreted as being indicative of a landslide. Parts of the houses that were on fill therefore were supported on pile foundations while front parts were on shallow foundations on stiff soil, so when the landslide re-activated, it pulled the props out in a spectacular manner and with little advance warning. Nuclear wars and landslides are not covered by ordinary homeowner insurance because they are regarded as unpredictable.

A No-No! Landslide Repair Method

As soon as it stops raining and a landslide stops moving, the natural tendency is to push the soil back where it belongs, which never works. Shearing displacements normally are accompanied by dilatant expansion as soil grains ride up and over one another, which in turn sucks in more water and permanently weakens the affected soil. A common field evidence is soil squeezing like toothpaste out from under the end of the landslide. It is the strength of the basal shearing layer that dictates the stability of the slide, and which is difficult to sample in triplicate for laboratory tests. More appropriate field tests are described later in this book.

When Landslides Stop

A landslide starts when the factor of safety, the ratio of the sum of resisting stresses to acting stresses, is 1.0. Shearing sucks in water and weakens soil, so sliding then will continue until it levels out sufficiently to compensate for the loss of strength. That also can be aided by drainage after it stops raining, in which case the pause is only temporary until it starts raining again. That usually occurs after a period of heavy rains that saturate the soil, make it heavier, and increase pore water pressure that reduces friction.

A landslide can continue to move downhill until the soil weight is reallocated sufficiently to compensate for the loss of shear strength in the basal zone, and sliding stops. Clay soils can regain some strength through *thixotropic setting*, which can postpone a renewal of sliding. The longer a landslide remains stopped, the more the soil shear strength may increase through drainage and aging. A landslide, therefore, can remain deceptively stable and clues such as landslide scarps can be obscured under vegetation until serious activity causes it to start up again. Common causes are adding load near the top, removing lateral support near the bottom, earthquakes, and saturation from septic drain fields.

Recognizing Landslides

Landslide scarps are obvious unless covered up. Dead trees and trees that are tilted in a downhill direction can be a clue, but trees also can become tilted from seeking sunlight. One of the more positive identifiers for a landslide is a stream that is pinched shut at the bottom of a hill and may pond water. Interruptions of sewer and water service can provide evidence that is difficult to ignore or overlook. At the first hint of a landslide, the gas company should be called, so it can install flexible connecting lines.

Not a Good Place for a Patio

A walkout basement is readily arranged on a hillside and is a popular way to save the cost of removal of excavated soil by pushing it out over the side and leveling it off to support an attractive patio until it starts to slide downhill.

1.13. Stopping a Landslide

Drainage

Removing water reduces soil weight and reduces pore water pressure, two important objectives that can increase the stability of sliding soil. However, there can be difficulties. An open trench in an active landslide is unlikely to remain open, and only short sections should be opened at one time, during a pause in sliding and in dry weather. Drains also can become a liability if sliding pinches or breaks a drain line.

An increasingly popular option is directional drilling that is controlled electronically at the cutting end and was developed in the petroleum industry. Installation usually is from the top down, with the pipe exiting at the lower end of the slide. Continuous hard-plastic drain pipes resist pinching or breaking, and are perforated to allow entry of water.

Structural Restraints: Piles, Stone Columns, and Retaining Walls

Piles and stone columns may be used to pin down a landslide, but a close spacing is required because of the large forces involved, and soil between the restraints may continue to slide. Piles must be firmly anchored at the bottom because of the large bending moment from soil pressure concentrated near the middle.

Conventional retaining walls are designed to resist *active state soil pressure,* where friction between soil grains acts to retain the soil and reduce lateral pressure. Stopping an active landslide requires resisting *passive state pressure* where soil is doing the pushing so friction is reversed. A bulldozer overcomes passive soil pressure unless it backs up, when pressure is relieved to the active state. In soil having an internal friction angle $\varphi = 20°$, the active pressure coefficient is $K_a = (1 - \sin \varphi)/(1 + \sin \varphi) = 0.5$. The passive pressure coefficient is the inverse, $K_p = 2.0$, so the ratio of passive to active $2/0.5 = 4$. That can call for a massive retaining wall.

Reinforced earth is massive, somewhat flexible, and thick enough to contribute to a large resistance to tilting or overturning. It was invented by a French engineer, Henri Vidal, after observing stabilizing effects from layers of pine needles on sand castles at the beach. A historical precedent is straw in the bricks, but the function of the straw is to mainly control shrinkage through uniform drying.

Reinforced earth now can be referred to as mechanically stabilized earth, or MSE. An MSE wall consists of layers of sandy soil between horizontal steel strips that are bolted at one end to wall panels. The strips are held in place by friction with the soil. The design follows conventional practice and is discussed in an introductory course in geotechnical engineering. Clay is not appropriate for use in an MSE wall because of poor drainage and a lower internal friction.

Chemical Stabilization

Atterberg limits are moisture contents that relate soil strength and behavior to its moisture content. The tests are discussed in the next chapter.

The *plastic limit* is the moisture content at which a soil changes from stiff and crumbly, to soft and remoldable. The moisture content of soil in the base of a landslide inevitably exceeds the soil plastic limit. The plastic limit can be sensitive and manipulated by mixing a clay soil with a few percent hydrated lime, $Ca(OH)_2$. A soil behavior can be changed from plastic to solid and crumbly by increasing its plastic limit. A simple field test for soil reactivity is described in Sec. 2.10.

Drilled Quicklime

Drilled lime, or boreholes filled with hydrated lime, is commonly used to stabilize expansive clay soil under a pavement without removing the pavement. The stabilization process is slow because of the slow rate of dissolution and migration of the lime. The process has been modified for stabilizing landslides in soils containing expansive clay minerals by substituting unslaked, pebble quicklime, CaO. Only the active or "pushing" zone of a landslide requires treatment, and can be identified using an analytical procedure called the *method of slices.*

Care is required in handling because of the caustic nature of the lime. The pebble size and shape are inherited from the crushed limestone that is used to make the lime, and reduces dusting. The spacing between quicklime columns is 10–12 times their diameter, and they *must* extend all of the way down through the bottom of the landslide:

1. Each 4- to 6-in. (100–150 mm) diameter boring is made with a continuous *flight auger* to extend into stable soil under the landslide.
2. A scoop-full of quicklime is dumped into the boring and workers stand back as the lime sinks to the bottom of any groundwater that has seeped into the boring and heats the water sufficiently to geyser any water out of the boring. The lime stays in the bottom of the boring.
3. Lime is poured into the boring to fill it to within about 0.5 m (1.5 ft) from the ground surface, and the remainder of the boring filled with soil. This is to reduce the likelihood of personal contact with the lime as it hydrates and expands upward in the boring.

The process has been successfully used many times, but requires care and expertise. Sliding usually immediately stops. Soil that is too soft to hold an open boring is penetrated with a long-shaft concrete vibrator, and the boring held open by pouring in lime.

Mechanisms: A hypothesis is as follows: (1) Drying action by the quicklime halts the landslide. (2) OH^- ions from the lime react with H^+ ions from expansive clay minerals to make the particles more negative. (3) Negative clay particles are linked by Ca^{++} ions released from the lime to flocculate the clay. (4) Over a period of years the high pH dissolves clay minerals and creates compounds that have been identified by X-ray diffraction as being the same as those in hydrated Portland cement, solidifying the soil into a permanent, rock-like hardness. Landslides stabilized by this method have remained stable for over 50 years.

1.14. Rock That Isn't There

Limestone caverns are hollowed out by infiltrating rainwater with dissolved carbon dioxide: $H_2O + CO_2 \rightarrow H_2CO_3$, which is slightly acidic. The water seeps downward in unsaturated soil until it encounters a groundwater table, where it becomes concentrated and slowly dissolves calcium carbonate in limestone: $CaCO_3 + H_2CO_3 \rightarrow Ca^{++} + CO_2\uparrow + H_2O$. The end product can be that which develops along the surface of the groundwater table. A cavern may deepen as drainage is affected, and a succession of caverns can form at different depth as nearby valleys erode deeper and lower the groundwater table. Different cavern levels usually are connected by vertical passages appropriately referred to by spelunkers as "glory holes."

Near-Surface Features

Far more common and more likely to affect shallow foundations are clay pockets filling near-surface voids in limestone. Voids are created by dissolution along vertical cracks, and overlying residual *A* and *B* topsoil and subsoil fill the cracks. This is shown in Fig. 1.7.

As suggested in the figure, such cracks and clay pockets may be missed by pattern drilling and exposed later by foundation excavations. The choices then may be to do some dental work by digging out the cavity and filling it with crushed rock, or if the cavity is too deep, by installing steel meth to support the rock and grout it solid. Another option is to move the structure.

Figure 1.7 Near-surface clay-filled solution cavities in limestone. Structural support may be required to bridge across the cavities. (Image source: *Geotechnical Engineering: Soil and Foundations Principles and Practice* by Richard L. Handy and Merlin G. Spangler. © 2007, McGraw-Hill Companies.)

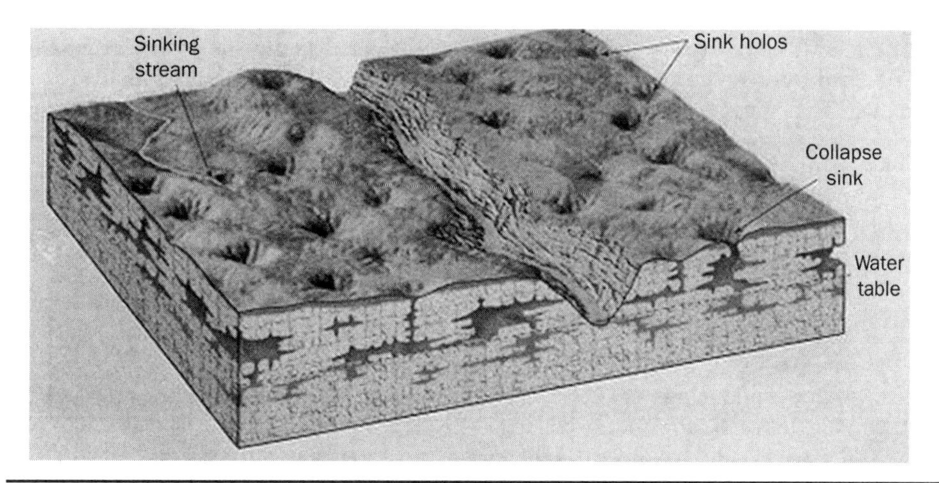

Figure 1.8 Karst topography, real holey ground caused by collapsed caverns in limestone. (Image source: Kentucky Geological Survey.)

Shallow Caverns and Sinks

Sinks can be an audacious clue to caverns as they are created by roof collapse. Streams flow in as they require no outlet. Sinks usually are brush-filled and can be collection points for old tires and refrigerators. Discretion is required before climbing down into a sink.

An abundance of sinks creates a *karst topography*, which is illustrated in Fig. 1.8. Extreme karst where sinks overlap can leave tall, rounded spires that are featured in traditional Chinese art.

Locating Underground Caverns

Probing and exploration borings are hit-or-miss, and shallow caverns have the least peripheral support so they may present the greatest potential for damage. A cavern obviously may only be under part of a building to cause damage, so test borings normally are located at outer corners. Collapse may occur in stages, or it can occur all at once with very little warning.

Electrical resistivity measurements are rapid, and the penetration depth depends on spacing between the electrodes. However, electrical current can take a detour over, under, or around a cavern so it is not readily detected. Electrical resistivity can be more useful for detecting clay pockets that are likely to be continuous along fractures.

An *engineering seismograph* is a small, portable instrument that is targeted for use in foundation engineering. Sound waves are generated with a sledge hammer hitting a steel ball or plate on the ground, and first-arrival vibrations are detected with *geophones. Refraction surveys* are most commonly used in foundation engineering, as sound waves follow a trapezoidal path down to rock, across through the surface of the rock, and back up to be detected at the ground surface.

Both resistance and seismic surveys can be set at a constant depth that equals or exceeds anticipated foundation depths by "walking" receiving points at a specific interval across an area.

Ground-penetrating radar is the latest innovation in an engineer's exploration toolkit, and the response is likely to be vertical and based on echoes. As the instrument is pulled over the ground surface at the speed of a walk, a continuous chart record is created that gives an immediate representation of rock surface contours. However, both wave penetration and velocity depend on the soil moisture content. The depth scale can vary by a factor of 2 or more, and will be nonlinear if there are variations in the soil moisture content.

Abandoned Mine Shafts and Tunnels

Vertical shafts that were mine entrances or used for ventilation pose obvious dangers, and animal bones are common at the bottom. An old shaft may only have been covered with a few boards when a mine was abandoned.

Any shallow depression in a mine area is suspect, but curiously the deeper the mine, the less likely that a collapse will cause a foundation problem. The reason is because rock involved in a tunnel roof collapse occupies more volume than it did prior to the collapse. Roof collapse can proceed upward in a process called *stoping* (with a long "o") until the entire volume of the original void is distributed in the rubble. This also can make a collapsed mine tunnel difficult to detect with drilling. One of the more reliable clues can be a sudden reduction in pump pressure in drilling mud that is circulated to bring up rock cuttings. Even more definitive can be pieces of saw-cut timber or chewing tobacco tins.

Tunneling Machines and the Rock That Isn't There

Tunneling was revolutionized and made safer by invention of horizontal tunneling machines by James Robbins in 1952. Prior to that, tunnels often were advanced by drill-and-blast procedures.

In a horizontal drill, the diameter of the tunnel is pushed ahead by anchors that expand laterally at the back of the machine to engage walls of the completed tunnel.

Preliminary exploration drilling from the ground surface is recommended because if the anchor fails to anchor, the machine may get stuck.

Case history. Vertical exploration borings were made every 50 ft (15 m) prior to boring a 6 ft (2 m) diameter tunnel at a depth of 90 ft (27 m) through limestone. The borings bracketed and missed a vertical seam filled with clay, and the machine found it. As there is not capability to back up, a vertical shaft was bored ahead of the machine and a tunnel cut to the machine so it could be pulled ahead to resume boring. The tunnel completion was delayed one year.

1.15. The Big Picture

In the 1930s a German meteorologist, Alfred Wegner, observed that the knee of Brazil could fit nicely into a recessed area on the west coast of Africa, and speculated that at one time they might have fit together. He examined fossil evidence, and discovered that fossils on both continents are identical up to a certain geological time and then went their separate ways. Wegener proposed a hypothesis he called *continental drift,* and he wrote books about it.

Wegener's unseemly idea apparently shocked leading geologists, who responded by organizing a symposium where they could take shots at it. Their main ammunition was the apparent lack of an energy source sufficient to break continents apart and push them around.

Wegener died on an expedition to Greenland before further evidence indicated that he was right. In the 1960s, radioactive dating of extinct volcanoes called *guyots* scattered across the bottom of the Atlantic Ocean revealed that the oldest parts of the Atlantic Ocean bottom are at outer edges, and it is breaking apart and separating along a line of active volcanoes called the *mid-Atlantic Ridge.* The driving mechanism is attributed to heat generated by radioactivity. Wegener's concept has evolved into *plate tectonics,* as plates can occur under oceans as well as under continents.

Mountain Ranges, Volcanoes, and Earthquakes

Compression and upward crumpling of a continental plate can occur where it overrides an ocean plate, as along the western edges of North and South America to create Rocky and Andes mountain ranges bordered by an offshore ocean deep.

Earthquakes are a product of stick-slip along plate contacts that also are avenues for emerging volcanoes. Active volcanoes outline the Pacific Ocean with a "rim of fire."

The Hawaiian Islands are in isolation, and their decreasing age from west to east can be explained by a plate drifting over a hot spot.

Mountain ranges slowly succumb to erosion, which can broaden their exposure into a granitic "shield area" such as that occupying much of eastern Canada. Shield areas retain intersecting vertical cracks occupied by streams, lakes, and beavers.

Soil Responses to Earthquakes

Ground shaking affects contact stresses and friction between grains of water-saturated, low-density sand, so grains may slip into adjacent voids and are separated by water to temporarily become *quicksand.* Unlike quicksand that can be attributed to a rising current of water, the condition is temporary because sand grains settle out. Hence, the process is more accurately called *liquefaction.*

Liquefaction of a foundation soil can take buildings down in a matter of minutes as they tilt and sink to equilibrium into the ground. Particularly vulnerable is "made ground" commonly created along bay margins by dumping sand into water without any attempt at compaction.

Prof. H. B. Seed (1922–1989) at the University of California, Berkeley, pioneered research into liquefaction, and developed correlations that are discussed later in this book. Liquefaction of sandy soil can be prevented by compaction, in particular deep vibratory compaction that gets there first.

Clay soils also can be sensitive to earthquake vibrations that can echo within a deposit, so consequences are less predictable. Various methods for *ground improvement* are discussed in the last chapter of this book.

Earthquake Recurrence Intervals

A common clue to past episodes of liquefaction is small cones of sand that are deposited at the ground surface by water that is displaced as sand settles out and seeps upward through tiny channels. This is most common on river floodplains because of an abundance of sand, and where the cones are likely to be covered up and preserved under deposits of clay. Radioactive dating of organic matter in the clay then can establish past earthquake dates and recurrence intervals. The intervals are too irregular to allow accurate predictions of future earthquakes.

The largest earthquakes on record in the United States were not in California, but occurred in 1812–1813 near a town of New Madrid in southeast Missouri. Ground rolls made it difficult to stand, and ground sinking following liquefaction probably contributed to the formation of Reelfoot Lake in nearby Tennessee. Parts of the Mississippi River were observed to temporarily flow backward. The earthquakes can be attributed to faults involved in earlier episodes of mountain building such as those that occurred to develop the Appalachian Mountains and the Ozark uplift, and created faults that remain subject to stick-slip.

Man-made earthquakes: The earth's crust remains stressed, and in the 1950s, a series of small earthquakes in Colorado were attributed to pumping water into deep wells because when pumping stopped, earthquakes stopped.

Hydraulic fracturing, or *fracking*, of rocks to increase petroleum production was discovered in the 1940s during a grouting experiment, when cement-water grout that was pumped at high pressure into the ground flooded an irate farmer's field on the other side of a hill. Fracking substitutes water for grout and contains quartz sand that injects to keep cracks propped open. The procedure was widely adopted to increase petroleum production, and decades later was combined with directional drilling to aid retrieval of oil from shale with an array of wells extending outward from a center vertical well. The collection process is far more efficient than the use of separate vertical wells, and horizontal boring along shale layers costs less than vertical drilling that must go down through layers of hard rock. Petroleum is carried along by water that is separated and remains contaminated, so it is put back where it started, deep into the ground in depleted oil wells. As the use of fracking has increased, so has the occurrence of earthquakes in petroleum production areas, indicating a revival of stick-slip, and earthquakes now are more common in Oklahoma than in California.

Conflicting possibilities: Perhaps frequent small slips along an existing fault will relieve stress so it does not build up and cause a large one. Or may small slips transfer stress to more firmly bonded parts of a fault, so when it does let go the earthquake will be larger? Perhaps time will tell.

Buildings and earthquakes: Tall buildings develop a natural frequency, as when they sway with wind, and the frequency is much lower than that generated during stick-slip in an earthquake. Deep foundations are stiffeners that can tend to increase natural frequency, but usually not to a level normally generated by an earthquake. Buildings may be supported on rollers that can minimize the transfer of stress. Particularly damaging can be where floors on one push-in walls on an adjacent building, so they crumple and collapse. Evaluations of earthquake susceptibility are a specialized area of foundation engineering.

1.16. The Walkabout

One attraction of geotechnical/foundation engineering is that it can provide a legitimate excuse to go onsite and get some fresh air, and a site investigation should not be left to the discretion of the driller whose main concerns are drilling and conducting field tests. Even a preliminary walkabout by a geotechnical engineer with local experience can reveal clues to soil behavior. Clues to landslides are discussed earlier in this chapter. Cracks and tuckpointing in existing buildings are clues to differential movements that may involve uneven settlement or uneven uplift depending on whether a foundation soil is compressing or expanding. Such observations are unlikely to be recorded in driller's logs. Nearby existing structures also can be photographed in a casual manner. Most geotechnical firms maintain files for future reference but they are likely to be confidential and remain the property of the owner.

Problems

1. Sand next to a granite source is about 25 percent quartz and 75 percent feldspar. In river and beach sands, the percentages are approximately reversed. Why? How may this affect the grain shapes and sizes and physical properties of the sand?
2. How did continental glaciation affect alluvial soils in the tropics?
3. Explain the occurrence of sand and gravel in high terraces bordering many river floodplains in the northern hemisphere. Is more than one terrace level possible? Why is the downstream gradient of many terrace remnants steeper than that of the modern floodplain? How might that affect prospecting for gravel?
4. In Fig. 1.3, find a partially clay-filled oxbow lake. How did it form and what is special about the clay?
5. Borings in the lower Mississippi River floodplain reveal a deep sand and gravel-filled canyon. Explain. Why is it significant for foundation engineering?
6. How may a leaky water faucet or broken sewer pipe affect a foundation on loess soil close to the loess source?
7. What do the development of Darwin's and Wegener's theories have in common, and how do they differ with regard to: (a) location and nature of observations that led to the theories; (b) sources of opposition; (c) areas of acceptance?

Further Reading

Greene, M. T., *Alfred Wegener: Science, Exploration, and the Theory of Continental Drift*, John Hopkins University Press, Baltimore, MD, 2015.

Handy, R. L., "Loess Distribution by Variable Winds," *Geological Society of American Bulletin*, 87:915–927, 1976.

Alfred Wegener (1880–1930), a scientist and meteorologist who proposed drifting of continents, initially ridiculed and now widely accepted. He died on a meteorological expedition while attempting to find and rescue fellow explorer on the Greenland icecap.

CHAPTER 2

Getting along with Classification

What Chaos Needs Is A Little Organization

2.1. A Hands-On Experience

Albert Atterberg (1846–1916) was a Swedish agricultural chemist, and he devised some simple soil tests that are widely used in engineering soil classifications. He discovered that when some moist clay was molded and remolded between the palms of his hands, as it dried out it suddenly would become crumbly. He weighed the crumbs, dried them in an oven, and weighed them again, and discovered a consistency that also varies from soil to soil. The moisture content for conversion to crumbly is the *plastic limit*, common designation *PL*. It is an *Atterberg limit*.

2.2. An Engineered Soil Moisture Content

The *dry weight* of soil stays constant as the amount of water varies, so the dry weight is used as the base for a percent moisture. Soil is the container. When defined on a *dry-weight basis*, a soil moisture content MC is

$$\mathrm{MC} = 100 \times (W_w - W_d)/W_d \tag{2.1}$$

where W_w is the total or wet weight, and W_d the weight after oven-drying. The oven temperature is standardized at 105°C.

Does a dry-weight definition really make much difference? If W_w for a sample of soil is 100 g and $W_d = 75$ g, then:

- Ordinary definition of MC that engineers regard as unworthy: $(100 - 75)/100 = 25\%$.
- Definition used in geotechnical engineering: $(100 - 75)/75 = 33\%$.

It makes a difference.

Question: Can a soil moisture content possibly exceed 100 percent?

Answer: Let $W_w = 100$ g and $W_d = 40$ g; MC = 100(100 – 40)/40 = 150%. (This actually can occur in some weathered volcanic soils subjected to daily rains that prevent clay minerals from crystallizing.)

2.3. Standardizing the Plastic Limit Test

The palm of the hand can be used to roll a sample of moist soil down into threads that are about 1/8 in. (3 mm) in diameter (see Fig. 2.1). The threads then are molded together and rolled out again, and the procedure repeated until the soil becomes crumbly. The crumbs are weighed, oven-dried, and weighed again to determine the moisture content at which the soil crumbles when rolled out into threads. That moisture content is the plastic limit, or PL.

The Plastic Limit in Engineering

The moisture content of soil in the base of a landslide exceeds the soil plastic limit, and sliding can be stopped if the moisture content is reduced below its plastic limit. The PL of a fine-grained soil involved in a bearing capacity failure often will exceed the soil plastic limit. On the other hand, that criterion does not apply to sand because it is *non-plastic*. The behavior of sand depends on friction between the grains and is influenced by dilatancy, an increase in volume as grains slide up and over one another.

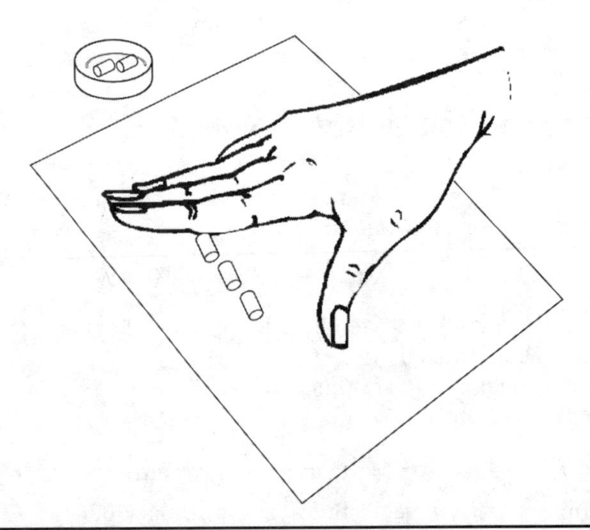

FIGURE 2.1 A hands-on approach to measuring a soil plastic limit. A modification uses a flat paper-covered board to do the rolling, and the paper is replaced after every test. (Image source: *Geotechnical Engineering: Soil and Foundations Principles and Practice* by Richard L. Handy and Merlin G. Spangler. © 2007, McGraw-Hill Companies.)

2.4. Going from Plastic and Remoldable to Liquid and Flowable

Atterberg modified the hands-on approach to define a soil *liquid limit*, or LL, the moisture content that can change it to a liquid. The transition is a bit more complicated because of *thixotropy*, a tendency for moist clay to set if it is not disturbed. The same process occurs in an ice-cream milk shake and can be conquered with a spoon or a straw. Atterberg spread some wet clayey soil in the palm of his hand and cut a groove with his finger, then tapped the back of his hand to observe if the groove closed. Tapping is required to overcome the thixotropy. Tests can be conducted at different moisture contents and the one that allows the groove to close with a particular number of taps that has been standardized at 25.

Standardizing the LL Test

The Atterberg definition seemed a little loose, so Harvard Professor Arthur Casagrande (1902–1981) standardized the test using a cam arrangement discovered in a notebook of Leonardo Da Vinci (1452–1519). As Leonardo found no particular use for his cam, it may be what contributed to the Mona Lisa smile.

Casagrande substituted a rounded brass cup for the palm of the hand and used a standardized tool to make the groove. The device is shown in Fig. 2.2, and is adjusted so one revolution of the crank lifts and drops the cup a distance of 10 mm. It impacts on a hard rubber block. The test is performed at several different moisture contents, and a graph is plotted so the moisture content can be picked off, which will allow the groove to close at 25 blows.

The Fall Cone Test

Casagrande's approach being a bit complicated, a simpler approach is by use of a cone penetration test that was developed in Sweden in 1915. As shown in Fig. 2.3, a smooth stainless steel cone with a 30° tip angle and a mass of 80 g (including the shaft) is lowered so the point touches the surface of a soil sample. The cone then is released and the penetration distance measured after 5 seconds. In a procedure similar to that used in Casagrande's test, the process is repeated at different soil moisture contents, and the results are graphed

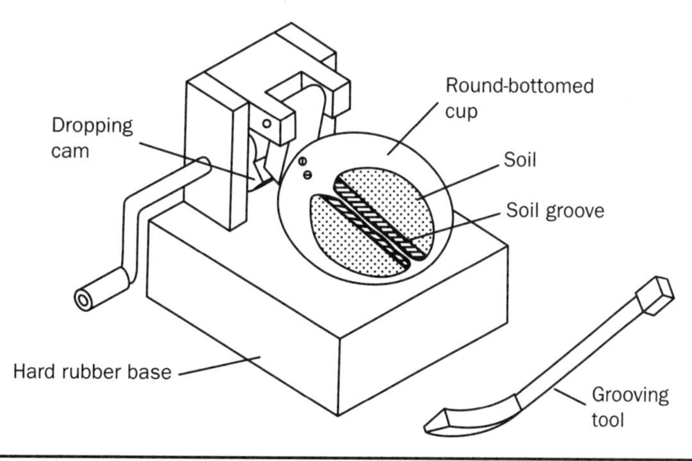

FIGURE 2.2 The Casagrande liquid limit device uses a lifting cam invented by Leonardo da Vinci. (Image source: *Geotechnical Engineering: Soil and Foundations Principles and Practice* by Richard L. Handy and Merlin G. Spangler. © 2007, McGraw Hill Companies.)

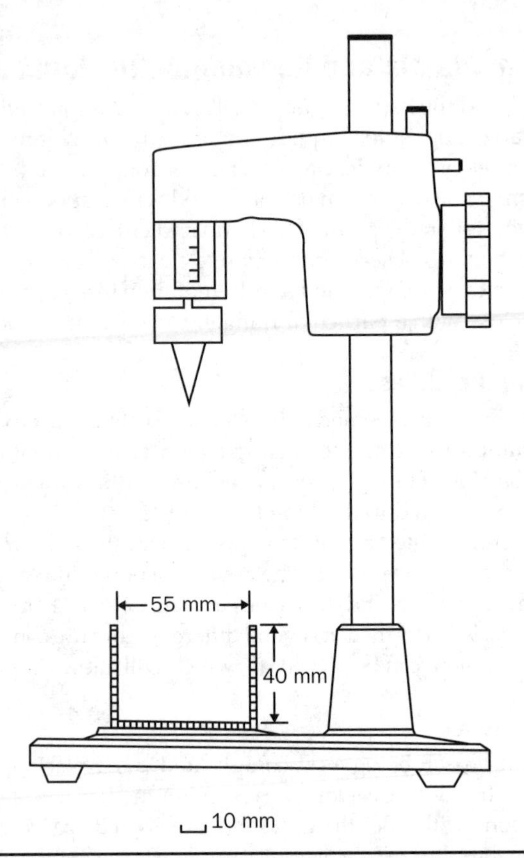

FIGURE 2.3 Diagram showing the fall cone test apparatus and standardized cup to hold the soil. (Image source: Courtesy of Prof. A. J. Lutenegger, Univ. of Mass., Amherst.)

to determine a moisture content at which the cone has penetrated 20 mm into the soil. Test results from both procedures are comparable and test variability is reduced about one-half with the cone test, and it has been adopted as an alternate standard method. Coincidentally, the cavity width from 20 mm penetration of a 39° cone is almost identical to the groove width in the Casagrande method. The fall cone method now is used in most laboratories.

2.5. The Plasticity Index

As the clay content and activity in a soil increase, the PL decreases and the LL increases, and so the numerical difference increases. It is the *plasticity index*, or *PI*. The plasticity index is not a percentage; it is a numerical difference between percentages, the range in moisture contents through which a soil behaves plastically:

$$\mathrm{PI} = \mathrm{LL} - \mathrm{PL} \tag{2.2}$$

The PI is high in expansive clay, lower in non-expansive clay, even lower in silt, and zero in sand. The test cannot be performed in sand and it is reported as *NP*, for nonplastic. Details of the tests are standardized in ASTM D43-18.

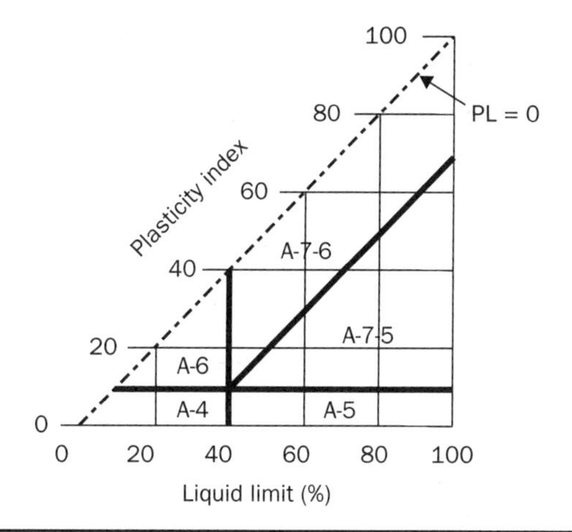

FIGURE 2.4 Chart to aid in classification of fine-grained soils in the AASHTO system. *Example:* PI = 45, LL = 60%; the soil is *A-7-6*, expansive clay. (Image source: *Geotechnical Engineering: Soil and Foundations Principles and Practice* by Richard L. Handy and Merlin G. Spangler. © 2007, McGraw-Hill Companies.)

2.6. Atterberg Limits in Soil Classification

Shortly after coming to the United States from his native Austria, Karl Terzaghi collaborated with Chester Hogentogler at the U.S. Bureau of Public Roads to devise a classification system based on Atterberg limits and now called the *AASHTO* system, having been adopted by the American Association of State Highway and Transportation Officials. Figure 2.4 shows a chart that can be used to classify fine-grained soils. The AASHTO system is mainly used in highway engineering but it can convey important information for foundation engineers. As "A" numbers go up, engineering qualities go down:

- *A-1:* Gravel and crushed stone. Can be compacted for road base courses and to support slab-on-grade foundations. Also employed in vertical "stone columns" and rammed into columns for *ground improvement*, discussed in Chap. 12.
- *A-2:* Sand. Weaker than A-1 but drains and if well-graded compacts well. Often used as a leveling layer under floors and foundations.
- *A-3:* There is no A-3. It is a break that separates good from marginal and bad soils for engineering uses.
- *A-4:* Silt. Finer and weaker than sand and includes low-density, wind-deposited loess that can collapse when saturated with water.
- *A-5:* Silty soil containing enough mica to make it springy and difficult to compact. Not common, usually a product of weathering of a metamorphic rock called *mica schist*.

- *A-6:* Intermediate clay. Hard when dry, soft when wet and usually moderately expansive: Shrinks when dry, expands when wet.
- *A-7-5:* Silty clay that can be expansive.
- *A-7-6:* Plastic clay that is expansive and avoided if possible. A small percentage of hydrated lime can change it to A-4.
- *O:* Organic soils, usually peat, and unacceptable as a foundation soil.

2.7. WWII and New Rules for Soil Classification

In 1942, at the request of the U.S. Army Corps of Engineers, Prof. Arthur Casagrande at Harvard University devised a soil classification for field use by the military. The system later was modified and adopted by both the Corps and the U.S. Bureau of Reclamation for general use, and is called the *Unified System.* Soil designations are by letters that are somewhat self-explanatory.

Unified System designations are summarized in Table 2.1. As with the AASHTO system, the usefulness of soils for engineering purposes decreases downward in the table. Estimates can be made without testing, and this is the system that is most commonly used in foundation engineering. A graph that can be used in classification is shown in Fig. 2.5.

CH normally means expansive clay that is particularly troublesome. *ML* includes wind-deposited loess soils that may collapse under their own weight if they become saturated with water. These soils are given special attention in Chap. 4.

G = gravel:	*Two subclasses:* **GW** if well-graded, meaning a broad coverage range of grain sizes so the soil can be compacted to a high strength and density. **GP** if poorly graded and less compactable. May either be "skip graded" or have a limited range in grain sizes.
S = sand:	50 percent or more sand. *Two subclasses:* **SW** if well-graded, having a wide range in grain sizes so it compacts to a high density. *(This characteristic is referred to by geologists as "poorly sorted.")* **SP** if poorly graded and less compactable. A limited range in grain sizes. *("Well-sorted.")*
M = silt*:	Based on both plasticity index and liquid limit: **ML: Lower range PI and LL $<$ 50.** **MH: Lower range PI and LL $>$ 50.**
C = clay:	**CL and CH: Same criteria but with higher range of PI.**
O = organic.	(Unacceptable)

* From German *moh* for silt.

Table 2.1 Soil Designations in the Unified Soil Classification System.

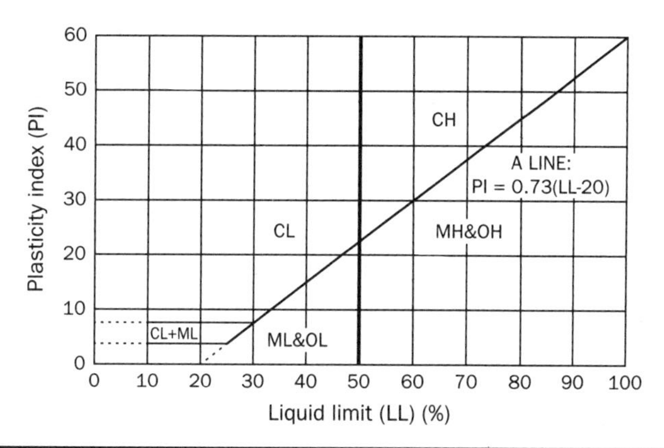

FIGURE 2.5 Chart to aid in classification of fine-grained soils according to the AASHTO system. *Example:* PI = 45, LL = 60%; the soil is *CH*, and probably is expansive clay.

2.8. Atterberg Limits and Criteria for Expansion

Figure 2.6 shows some criteria developed by F. H. Chen (1912–1999), a prominent consulting engineer. Chen's criteria were developed from laboratory measurements of undisturbed soil samples during wetting, and separate influences from floor versus foundation loads. As with any empirical correlation, results should be regarded as estimates or approximations.

Example: The soil cited in the caption for Fig. 2.4 has *PI* = *45*, *LL* = *60%*. It classifies as *CH* or *A-7-6*. According to the Chen criteria, is it likely to be expansive?

Answer: According to the charts in Fig. 2.6, the clay should be highly expansive under a floor load and moderately expansive under a foundation load. The expansion potential can be measured by introducing water to an undisturbed sample of soil in a consolidation test.

2.9. Kinds of Clay Minerals

Prior to the use of X-ray diffraction all clay was assumed to be kaolinite. Kaolinite is named after a village in China and its pure form is used to make porcelain. X-ray diffraction revealed that even tiny clay particles are crystals and their mineralogy is complicated. Tests of clays that were allowed to equilibrate at different relative humidities then revealed that some clay minerals take varying amounts of water between crystal layers so they can shrink and expand. Engineering tests revealed that their expansion can create enough pressure to lift a building foundation.

A Layered Crystal Structure

The kaolinite crystal structure has two *sheets* that share atoms to make a layer shown diagrammatically in Fig. 2.7. Layers are held together by *hydrogen bonds,* hydrogen atoms that are stripped of their electrons so they are protons, tiny and intense positive

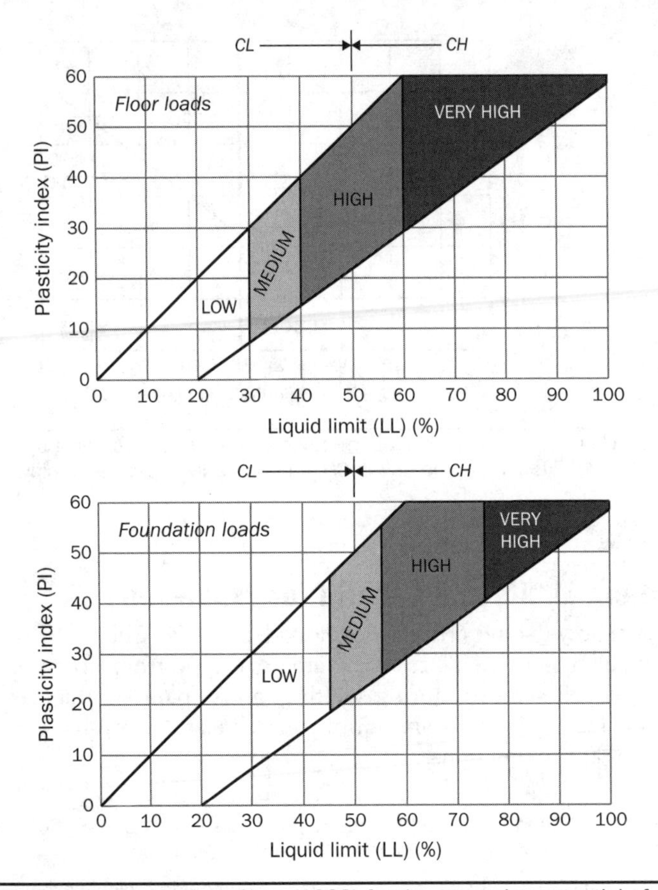

Figure 2.6 Criteria developed by F. H. Chen (1988) for the expansion potential of expansive clays used to support for floors and foundations. (Image source: From Handy (2002). With permission from *Geotechnical News*.)

charges. Hydrogen bonds are weak enough that sheets can be split apart and strong enough that they are not separated by water.

An Expansive Crystal Structure

Expansive clay minerals have three sheets to a layer, shown in the lower part of Fig. 2.7, so there is no interlayer hydrogen bonding. However, the additional oxygen ions contribute to a net negative electrical charge. It is balanced by positive ions, *cations* (pronounced cat-ions) held between the layers. It is the interlayer positive ions that can control expansion of the clay.

The lower part of Fig. 2.7 shows interlayer potassium (K^+) that is a close fit and prevents layer separation. The mineral is *mica*, which is not expansive and can exist in large crystals. Mica has the structure of a clay mineral but is not a clay mineral. The crystals are too big to be clay.

If potassium is removed by weathering, a common replacement is calcium (Ca^{++}) that allows a controlled expansion from taking in 0, 1, 2, or 3 diameters of water

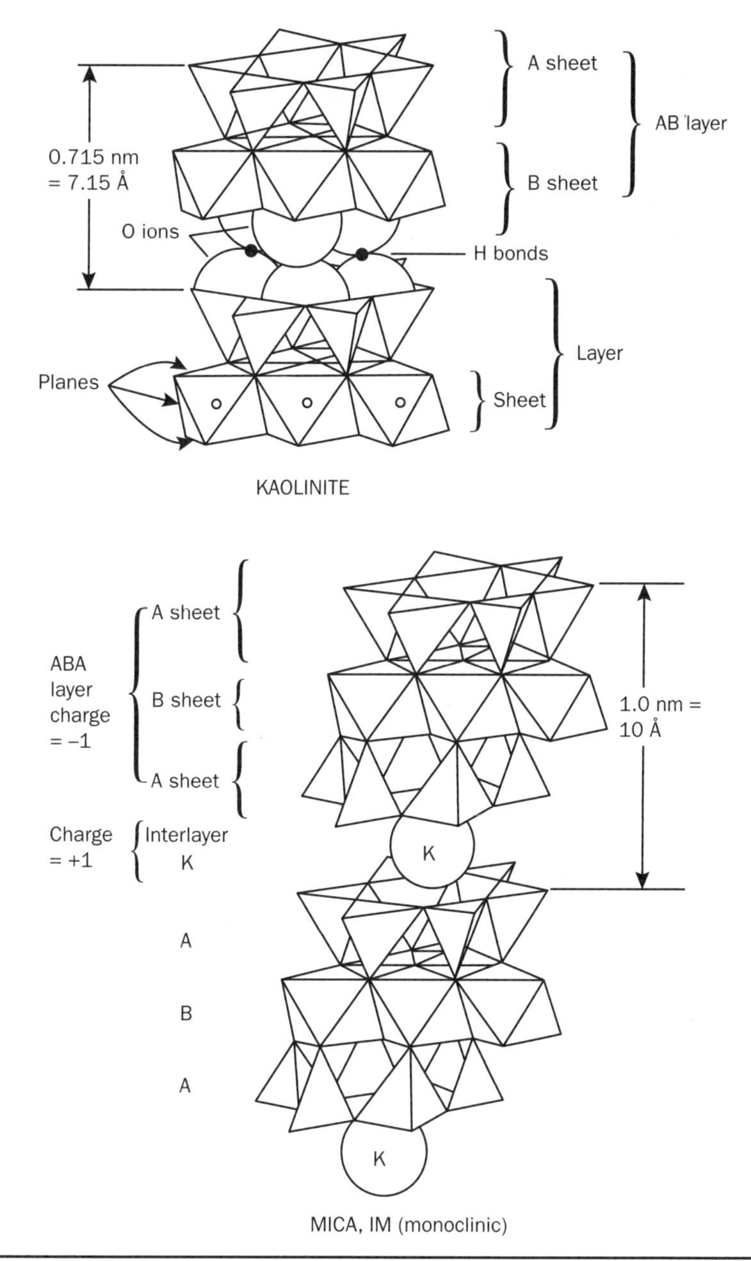

FIGURE 2.7 Representations of crystal structures of two- and three-sheet silicate minerals that include clay minerals. Points of the triangles represent centers of oxygen atoms. In the lower figure, potassium K^+ ions hold mica layers together so it is not expansive. (Image source: *Geotechnical Engineering: Soil and Foundations Principles and Practice* by Richard L. Handy and Merlin G. Spangler. © 2007, McGraw-Hill Companies.)

molecules. Calcium is the most common interlayer cation in expansive clay in soils. A clay mineral can be tricked into taking in more calcium, which stabilizes the clay for engineering uses. This process is discussed in Chap. 4.

Going Tribal

The chemical composition of expansive clay minerals is variable so they are called *montmorillonite group minerals,* named after a village where first defined in France. A more recent name is *smectite.*

When Sodium, Na^+, Replaces Calcium, Ca^{++}

The reduced positive charge of sodium can allow expansive clay to expand indefinitely, so the substitution is used to disperse soils for a grain size analysis. The substitution is forced if the calcium that is released is trapped in another molecule. The process is used to remove calcium from "hard" water with water softeners.

Drilling Mud

Commercial *bentonite* is relatively pure expansive clay that is weathered volcanic ash. When a lava is blown into the air, droplets cool so rapidly there is no crystal structure; they are glass, which in turn readily weathers into expansive clay.

The high water-holding capacity of expanded bentonite that makes it useful for drilling, being pumped down the drill pipe to come up around the outside and bring up rock fragments. If drilling is interrupted, the mud "sets" and keeps suspended drill cuttings from settling out and jamming the drill. The property, called *thixotropy,* also occurs in milkshakes and is disrupted by stirring with a straw. Bentonite also is improperly used for *mud jacking* to lift sagging pavements because it lets them back down when the mud dries out.

2.10. A Hands-On Test for Expansive Clay

Only a few percent hydrated lime, $Ca(OH)_2$, can increase the plastic limit of expansive clay so it exceeds the existing moisture content. The clay almost instantly becomes crumbly. A possible explanation is that OH^- ions from lime combine with pull H^+ ions from the clay to make H_2O. The clay particles are more negative and are linked by the calcium Ca^{++} ions released from the lime. Scanning electron micrographs indicate an open "cardhouse" structure that holds free water and is created by edge-to-face bonding between clay platelets.

Field Test

"Pickling lime," $Ca(OH)_2$, can be purchased from the pickling section of a grocery store. A small amount of lime is added to clay *at its existing moisture content* and sealed in a plastic bag, and kneaded with the fingers to determine if the clay becomes crumbly. It is not a difficult test. If the clay suddenly stiffens and becomes crumbly, it most likely is expansive.

2.11. Some Clues to Expansive Clay

Faulted and shifting pavements and sidewalks, cracks in walls, tuckpointing that has cracked in brick walls, wide strips of caulking, fireplace chimneys that are separated or tilted away from a house, ground cracks that may have to be looked for on hands and knees, nervous homeowners, and For Sale signs are some clues to expansive clays. Damages from expansive clays, like those from landslides, earthquakes, and foundation failure, usually are not covered by home insurance. They are too common, too difficult, too expensive.

2.12. Measuring Soil Particle Sizes

Another important part of soil classification is grain sizes that go all of the way from clay size up to silt, sand, gravel, cobbles, boulders, hills, and mountains. Cobbles and boulders are obvious but smaller sizes are separated and measured by passing through a stack or "nest" of sieves. Clay particles are separated and measured from their sedimentation rates in water.

Sieves are arranged with the coarsest at the top. Clay must first be removed from the soil so grains do not stick together, and so after being soaked in water, a sample of soil is vigorously stirred with a dispersing agent that substitutes sodium for other ions held on the clay. The cleaned soil is dried, weighed, and separated into size fractions by sieving. The amount retained on each sieve is expressed as a percentage and can be shown with a bar diagram as at the top of Fig. 2.8. The peak is a statistical *mode.* If there are two peaks, the soil is *bimodal.* There are two peaks in Fig. 2.8.

The more usual way to graph the same data is in a *particle size accumulation curve,* shown in the lower part of Fig. 2.8. The data are the same but the bars are stacked as shown in the figure. The average, or *median,* grain size then can be directly read from 50 percent line on the graph.

Statistical Interpretation

The bar heights in the top of Fig. 2.8 often define a *Gaussian distribution* that reflects a randomness in deposition, as a river velocity varies as it is depositing a layer of sand.

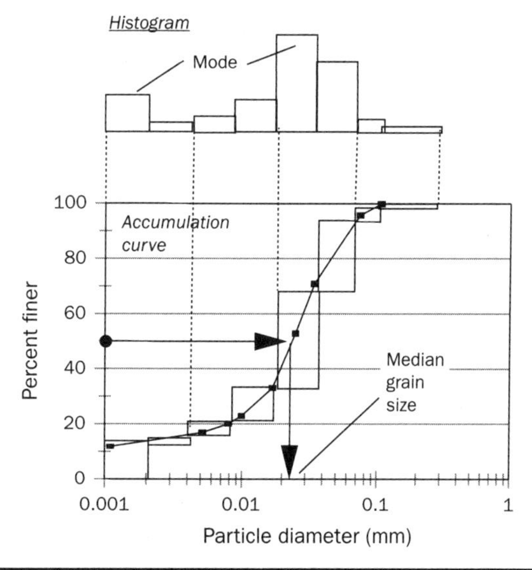

FIGURE 2.8 A *particle size distribution diagram* in relation to a statistical *histogram.* (Image source: *Geotechnical Engineering: Soil and Foundations Principles and Practice* by Richard L. Handy and Merlin G. Spangler. © 2007, McGraw-Hill Companies.)

The graphs in Fig. 2.8 are for wind-deposited silt or *loess* and reflect a variable wind velocity.

In the upper part of the graph in Fig. 2.8, the height of the bar at the left does not fit. It shows that there is a different mode for the clay. It can be explained if the clay was not acting as individual particles but was adhering to coarser silt particles. That is confirmed by microscopic examinations. The clay also is expansive so it can glue silt grains together so they can stand in high cliffs until wet with water.

Defining Clay Size

Graphs such as that shown at the top of Fig. 2.8 show a fairly consistent break at about 0.002 mm grain size, suggesting a change from non-clay to clay mineralogy, and that size is regarded as suitable size limit for clay. An earlier definition was set at 0.005 mm and was based on the resolution of a microscope in the U.S. Bureau of Soils (now the USDA). The 0.005 mm definition for the most part has been retained in engineering, not because it is accurate but for reasons that are suggested later.

2.13. Particle Sizes Determined from Sedimentation Rates in Water

George Stokes (1819–1903) was an Irish physicist and mathematician who specialized in fluid dynamics. He observed that large particles in water settle out faster than small ones, and derived an equation for the settlement of spherical particles. The equation now is referred to as *Stokes' Law* (Fig. 2.9). The relationship between settling velocity and the square of the particle diameter supports the use of a logarithmic scale for graphing diameters. As soil particles are not spheres, grain sizes measured in this way are "equivalent diameters."

Performing a Sedimentation Test

A representative sample of soil is soaked in water with a small amount of chemical dispersing agent, usually sodium metaphosphate (a water softener). The soil then is agitated with a stirring device such as shown in Fig. 2.10(A), and the soil-water mix then is diluted to 1000 ml with distilled water. The mixture is stirred to start the sedimentation process, and the cylinder is held at a constant temperature environment to maintain a constant water viscosity. However, it should be noted that the viscosity also depends on (a) volume displacement and (b) electrical interactions between suspended particles.

Stokes' Law for the rate of settlement of spherical particles. The law was modified by Einstein to account for fluid displacement by the particles, and other corrections have been made for electrical repulsions that can keep fine "colloidal" clay from settling out.

$$V_s = \frac{g(\rho_s - \rho_w)d^2}{18\mu} \tag{2.3}$$

where

V_s = terminal settling velocity of the solid particle
g = gravitational acceleration
ρ_s = density of settling particle
ρ_w = density of water
d = diameter of particle
μ = diameter viscosity

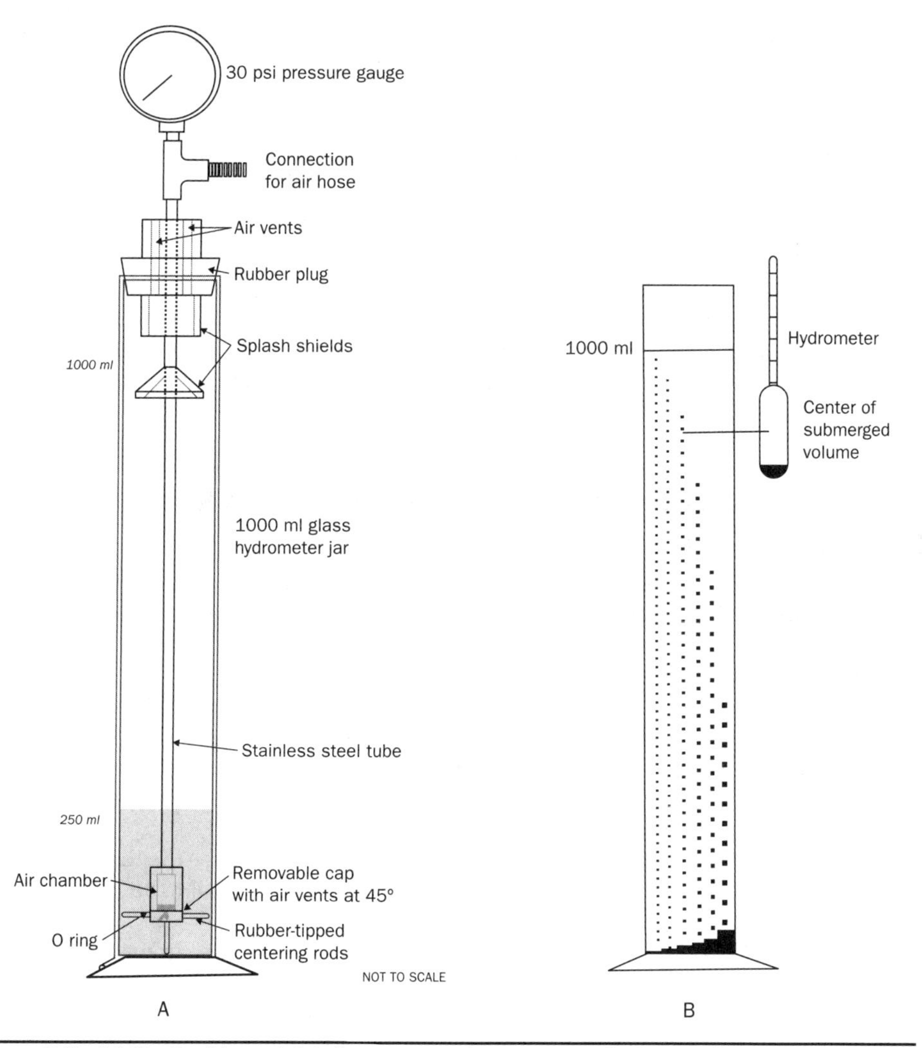

FIGURE 2.9 (a) Chu-Davidson air-jet disperser used to prepare soils for a sedimentation analysis. (b) Concept of the sedimentation analysis, as large particles settle faster and sink below the level of the measurement. (Image source: Suggested by Prof. T. Y. Chu.)

The concept involved in sampling is illustrated in Fig. 2.10(B): After a certain period of time, particles larger than a specific diameter will settle a calculated distance so they will not be included in a sample of the suspension. The sample therefore can be dried and weighed, and the residue can be divided by the initial weight to give a percent finer.

A Substitute Measurement

Soil particles that are fine enough to still remain in suspension will affect the density and unit weight of the suspension, which can be measured with a hydrometer, a glass float that is weighted to remain upright and has a stem that projects above the water and can be calibrated to read g/cm^3 of soil in suspension. Because of the speed and convenience, this is the preferred instrument in engineering. Sources of error include an averaging effect since the density measurement crosses a range of depths and mixing at shallow depths occurs every time the hydrometer is inserted to take a reading.

Defining Clay Size

It may be coincidental, but a factor that may influence the definition of clay size is because a hydrometer determination of the 0.005 mm equivalent grain size can be made in a working day, and the 0.002 mm measurement comes in the middle of the night. It still can be estimated by interpolating a particle size accumulation curve, and typically is about 80 percent of the 0.005 mm content.

2.14. Some Soil Characteristics Related to Grain Size Distribution Curves

The curve at the upper left side of Fig. 2.10 shows the distribution of particle sizes for clay sampled from an oxbow lake. About two-thirds of the particles are smaller than 0.001 mm. That is a unique feature of this kind of clay, and a strong indicator is that it is expansive and remained trapped in the lake.

The steep graph for loess and the steeper graph for river sand can be explained by narrow ranges in the respective fluid velocities when the particles were being sorted out and deposited. A relatively uniform grain size maintains large pore sizes and a low compacted density and is considered *poorly graded* for engineering applications. Sand and silt are separated by the difference in settlement rates.

Particle sizes in the glacial till are far more variable because they are not separated as they are released from melting ice. Engineers consider the soil to be *well graded* because the particles fit together when compacted.

Gravel deposits (not shown) usually are *bimodal* because gravel is deposited first from rapidly flowing water, and sand is deposited later and fills in the voids. The two sizes are readily separated so sand goes in one pile and gravel in another, and then proportioned to make a dense concrete.

2.15. Defining Size Grades

Gravel/Sand

A separation between sand and gravel at 2.0 mm is convenient and corresponds to a standard sieve size. Coincidentally gravel usually is rounded pieces of rock that combine minerals, and sand particles are more likely to be single minerals.

Sand/Silt

Sand grains, whether deposited from wind or from water, tend to bounce so they accumulate in ripples and dunes. Silt grains are more likely to stay put and if available can accumulate in thick deposits. A natural boundary between sand and silt sizes is confirmed by the upper part of the loess curve in Fig. 2.10.

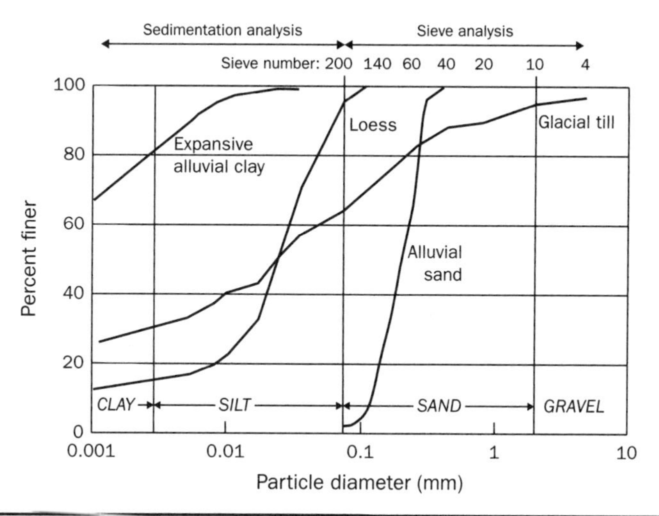

Figure 2.10 Examples of particle size gradation curves. (Image source: Modified from Geotechnical *Engineering: Soil and Foundations Principles and Practice* by Richard L. Handy and Merlin G. Spangler. © 2007, McGraw-Hill Companies.)

Clay and Silt

As previously mentioned, the distinction between clay and silt is mainly a matter of their mineralogy, either clay minerals or non-clay minerals such as quartz.

Expansive versus Non-expansive Clay

The two classes of clays are more accurately distinguished on the basis of their Atterberg limits, which therefore are incorporated into engineering classifications.

Salt versus Fresh Water Clay Deposits

Clay deposited in salt water is *flocculated* by the salt so it rapidly settles out in a delta. Clay deposited in a fresh-water lake is not flocculated and tends to accumulate as a milky sediment in the middle of the lake.

Problems

1. Use an available reference to classify the following soils in the Unified and AASHTO Systems:
   a. 95% passing No. 200 sieve, LL = 38, PI = 10.
   b. 10% passing No. 200 sieve, 85% passing No. 4 sieve, LL = 35, PI = 14.
   c. 2% passing No. 200 sieve, 9% passing No. 4 sieve, LL n.d., PI n.d.
   d. 88% passing No. 200 sieve, LL = 55, PI = 30.
2. Which, if any, of the above soils are likely to be expansive?

3. What is the basis for considering a soil to be "well-graded"? How may that influence the density and strength achieved by compaction?
4. With the addition of 2 percent hydrated lime, a soil LL goes from 68 to 60 and the PL from 43 to 23. How does that affect the soil classification? Why is it called "lime stabilization"?
5. Apply the Chen criteria to estimate the expansive character of the treated and untreated soil in the previous problem under floor and foundation loads.
6. The tests described in Problem 4 were repeated with 2 percent agricultural lime, $CaCO_3$, and the LL went from 61 to 59, the PL from 26 to 25. How did that affect the soil classifications? Why is $Ca(OH)_2$ so much more effective?

Further Reading

Chen, R. H., *Foundations on Expansive Soils,* Elsevier, Amsterdam, 453 pp, 1988.

Handy, R. L., "Geology, Soil Science and the Other Expansive Clays," *Geotechnical News,* 20:1, 2002.

Ho, C., and Handy, R. L., "Characteristics of Lime Retention by Montmorillonitic Clays," *Highway Research Record* 29, 55–69, 1963.

Albert Atterberg (1846–1916), Swedish agricultural chemist whose simple hands-on tests have become the basis for engineering soil classifications.

CHAPTER 3

Foundation Settlement

Original Plans for the Leaning Tower Probably Omitted the Part about the Leaning

3.1. Castles and Cathedrals

Medieval castles were built slowly, at least in part because of time required to squeeze tax money out of the peasants. Large, heavy stone blocks were used for the foundation layer, and progressively smaller stone blocks were placed on top. This arrangement was quite clever because the amount of settlement also is proportional to the logarithm of pressure, so early layers mattered the most. Necessary adjustments then could be made as layers stacked up, and medieval towers that are still standing almost inevitably have some kinks that were not part of the design. Castle designs also incorporated interconnecting walls for better coverage with bow and arrow and for bracing. It all came to naught following the introduction of gunpowder.

Cathedrals

High, vaulted ceilings of cathedrals point to the heavens but they also push outward on the supporting walls so it was not unusual for ceilings to fall in and interrupt the services. Natural arches, the Roman Pantheon, the St. Louis Arch, and the Innuit igloo are more stable because they follow the inverted shape of a chain held at the ends and there is no bending moment. The famous architect Christopher Wren studied the problem with scale models and suggested using long iron rods to hold the sides together. However, the usual solution was some flying buttresses to brace against exterior columns or walls and provide perches for gargoyles to ward off evil and keep an eye on the collection plates.

3.2. A Scientific Approach to Foundation Settlement

In the 1920s, Karl Terzaghi built a laboratory device to simulate foundation pressures on soils and opened up a new avenue in foundation engineering. No longer would it

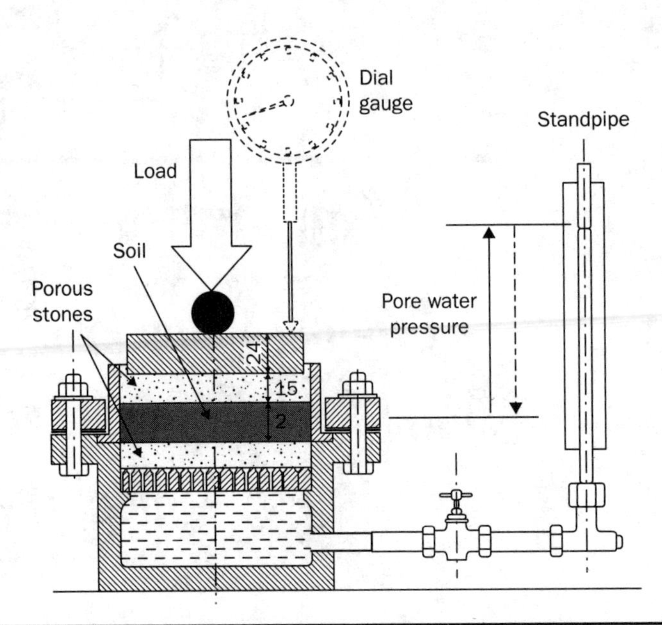

FIGURE 3.1 Terzaghi's consolidometer (dial gauge added for clarity). Modern instruments usually omit the standpipe, but it was critical for development of a theory of effective stress. (Image source: Modified from Terzaghi (1925).)

be necessary to build something and hope for the best, not if the amount and rate of settlement can be predicted.

The Test

A soil consolidation test is performed by incrementally loading a soil specimen that is held between two porous stones so that excess pore water can drain out. A diagram of Terzaghi's instrument is shown in Fig. 3.1. The soil is saturated with water to remove influences from capillary attractions, then incrementally loaded. After each load is applied, time is allowed for water to escape, indicated by water rising in the standpipe. Times were measured and recorded, and indicated when compression was complete. Twenty-four hours was found to be sufficient and is a convenient time for applying the next load increment. The procedure can be repeated until loading covers an anticipated field application.

As shown in Fig. 3.2, a small amount of pressure can be required to create a response that is proportional to the logarithm of the pressure, so it is convenient for each new load to double the previous load. (The *void ratio* uses the volume of solids as a reference, and is the calculated volume of voids divided by the volume of solids.)

A Eureka Moment!

Water rising in the small standpipe in Fig. 3.1 provides a measure of water pressure in pores in the soil, and water pressure pushes soil grains apart and reduce friction at the grain contacts. Thus was born the concept of *effective stress.* It is a valuable spin-off because it is relevant to landslides, which prefer wet weather, and to foundation bearing capacity failures that tend to occur during first loading when pore pressures are

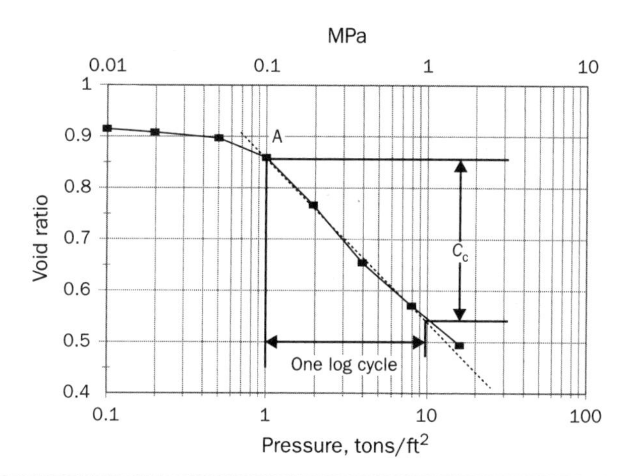

FIGURE 3.2 A linear semilogarithmic relationship between consolidation and pressure defines a *compression index*, C_c, that is used in settlement calculations. (Image source: *Geotechnical Engineering: Soil and Foundations Principles and Practice* by Richard L. Handy and Merlin G. Spangler. © 2007, McGraw-Hill Companies.)

highest. That observation has led to Terzaghi being acknowledged as the "father" of modern soil mechanics.

There was a drawback to the concept, which is that actual contacts between soil grains can't be occupied by water. Terzaghi assumed that the true contact areas are small and the error is on the safe side for design. Another factor has been recognized, which is clay mineral surfaces are stopping points for water molecules that can transmit pore water pressure.

> *The adhesion theory for sliding friction.* Terzaghi suggested that frictional resistance to sliding might be explained by temporary chemical bonding between small contact areas at tips of surface asperities: As contact pressure increases, asperities become deformed so the contact areas increase and so does friction. When pressure is released, friction between hard materials is lost due to elastic rebound. Contact areas between soft materials are more likely to be retained as cohesion. Terzaghi's concept has gone full circle back into mechanical engineering to explain the function of lubricants to keep surfaces separated (Bowden and Tabor, 1950).

3.3. Influence of Time

Figure 3.3 shows results from a typical consolidation test. Compression starts as soon as a load is applied. Note that time is graphed with a logarithmic scale. Three time-related stages of compression can be defined:

1. A brief first stage, which in this test lasted about 3 minutes, is typical, and can be attributed to compression required to restore the stress condition that existed

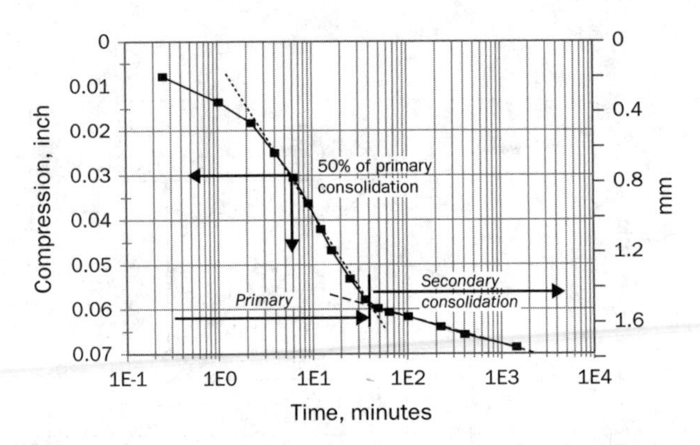

FIGURE 3.3 Representative consolidation test data after application of a single load increment. Primary consolidation is attributed to time required to squeeze water out of the soil, and then the mechanism changes.

in the field. That will include sufficient vertical pressure to re-create field horizontal stress, referred to as a K_o *condition*. Additional compression may simulate elastic compression and be recoverable. That can be measured by removing and re-applying a compression load.

2. *Primary consolidation* is revealed by the steeper slope of the graph, which depends on the kind of soil, slow for clay, faster for silt, and almost instantaneous for sand. Terzaghi related the slope to the soil *hydraulic conductivity:* The smaller the pipe, the more restricted the rate of flow.

   Layer thickness. In addition, the longer the pipe, the larger the resistance to flow. In a consolidating layer of soil, the escape route for water normally is up or down or in both directions depending on the nature of the boundary material. If one exit is sealed under a large foundation, the rate of settlement will be much slower. It can be improved with a layer of sand under a floor or foundation and peripheral tile drains. In theory that will increase the rates of settlement by a factor of 4.

3. *Secondary consolidation* occurs at a slower rate and does not depend on drainage distance is not controlled by excess pore water pressure. The rate may involve time-related shifts in weak chemical bonds between clay mineral particles, positive ions (cations), and adsorbed water, and is influenced by the kinds of interlayer cations. Different mechanisms make it difficult to link rates of primary and secondary consolidation, and the rate of secondary consolidation should not depend on the soil layer thickness as it affects the drainage distance.

Time-dependent particle bonding also may be the mechanism involved in slow downhill *soil creep* that occurs in clayey soils and can tilt small trees so they grow with curved trunks.

3.4. Amount of Settlement

The total amount of settlement can be calculated using the compression index, C_c, as shown in Fig. 3.2. The index represents *the change in void ratio resulting from consolidation across a one-log cycle*, and is readily determined from the steepest part of a graph as shown in Fig. 3.2. A linear relationship is described by the following empirical equation:

$$e_1 = e_0 - C_c \log (P_1/P_0) \tag{3.1}$$

A compression index strictly applies *only* to primary consolidation. Sometimes 24-hour settlement readings are used to define *e* values. As shown by the last data point in Fig. 3.2, it can include influences from secondary consolidation so it may overestimate primary consolidation settlement.

Example calculation using Eq. (3.1): In the test in Fig. 3.3, in the consolidating stage the change in void ratio in one-log cycle is $C_c = 0.85 - 0.55 = 0.30$. The preconsolidation pressure is estimated to be approximately 0.8 Tsf (0.08 MPa), and $e_0 = 0.91$. If the additional vertical stress created in the soil by an anticipated foundation load is 4.5 Tsf (0.45 MPa), what is the anticipated change in the void ratio?

Answer: $e_1 = e_0 - C_c \log (P_1/P_0) = 0.91 - 0.30 \log(4.5/0.8) = 0.68$.

Void Ratio and Settlement

The definition of a soil *void ratio* can lead to a simple expression between the amount of compression *S* of a consolidating layer of soil and the layer thickness *H*:

$$S = H\frac{e_0 - e_1}{e_0 + 1} \tag{3.2}$$

This derivation is included in many geotechnical engineering textbooks.

Calculating a Void Ratio

The total weight of a soil sample divided by its volume gives a *wet unit weight*. The sample then is oven-dried at a standardized temperature and the sample re-weighed to give an *oven-dry weight*. By convention the soil moisture content is the weight of the water lost divided by the *dry weight* of the soil and expressed as a percentage.

Example calculation of soil moisture content: A total weight is 10 and the dry weight is 5, both in the same units. The moisture content is calculated as MC = (10 – 5)/5 = 1.0 = 100%. Geotechnical engineering has a different outlook.

Weights must be translated into volumes of the solid fraction and of the voids in order to calculate a void ratio. The volume of solids depends on the specific gravity of the minerals that usually is in the range 2.65–2.70, the higher value being more representative of clay after oven-drying. An example calculation of void ratio with a wet unit weight of 120, a moisture content of 20 percent, and a mineral specific gravity of 2.7 is shown in Table 3.1.

Volume × Solid unit weight = Weight			
	Volume	Unit weight	Weight
Air	–	–	0
Water	–	62.4	***20****
Solids	0.60	2.7(62.4)	100
Total	1.00	–	***120****
** Measured. Then volume of voids = 1.00 – 0.60 = 0.40*			
Void ratio = volume of voids/volume of solids = 0.40/0.60 = 0.67.			

TABLE 3.1 Example Calculation of a Void Ratio

3.5. Overconsolidation and the Compression Index

Equation (3.1) only applies to normal consolidation, and most soils are overconsolidated. Even temporary lowering of a groundwater table during a 500-year drought will cause the soil to become overconsolidated after the groundwater level comes back up. The equation presents a valuable concept, but for better accuracy the void ratios used in settlement calculations should be based on actual measurements.

Example calculations of settlement: The pressure at the middle of a layer of soft clay soil is estimated to be 0.4 Tsf and a slab-on-grade foundation load will increase it to 1 Tsf. If the clay layer thickness is 5 ft (1.5 m), compare settlement (a) from data in Fig. 3.2 and (b) from Eq. (3.2).

Answer: (a) From the figure,

$e_o = 0.9$ and $e_1 = 0.86$.

Substituting in Eq. (3.2) gives:

$$S = H\frac{e_0 - e_1}{e_0 + 1} = 5(0.9 - 0.86)/(1 + 0.86) = 0.11 \text{ ft} = 1.3 \text{ in.} = \underline{33 \text{ mm}}.$$

(b) From Eq. (3.1), $e_1 = e_o - C_c \log (P_1/P_o) = 0.9 - 0.3 \log(1/0.4) = 0.9 - 0.12 = 0.78$.

Substituting in Eq. (3.2) gives:

$$S = H\frac{e_0 - e_1}{e_0 + 1} = 5\,(0.9 - 0.78)/(0.9 + 0.78) = 0.36 \text{ ft} = 4.3 \text{ in.} = \underline{109 \text{ mm!}}$$

Equation (3.1) does not take into account the overconsolidation.

3.6. Consolidation Rate

Darcy's Law for the rate of flow through a pipe is

$$q = kia \tag{3.3}$$

where k is the hydraulic conductivity (permeability), a is the cross-sectional area of a pipe, and i is the hydraulic gradient that is the reduction in pressure per unit length

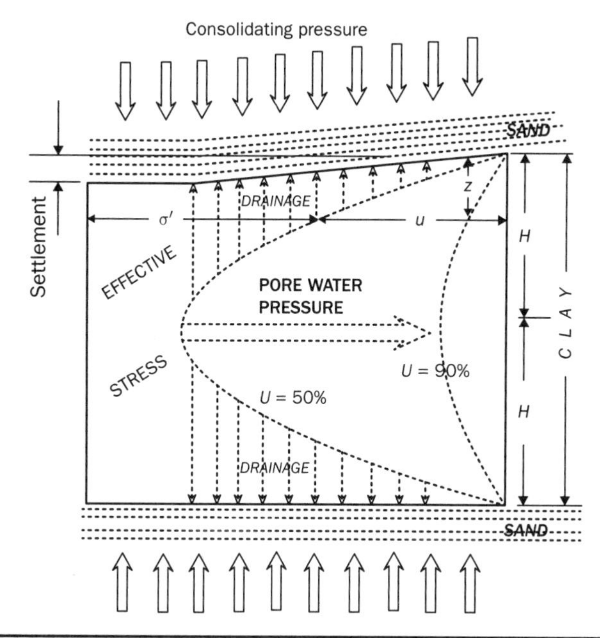

FIGURE 3.4 Theoretical distribution of pore water pressure and velocity for fluid flow. (Image source: *Geotechnical Engineering: Soil and Foundations Principles and Practice* by Richard L. Handy and Merlin G. Spangler. © 2007, McGraw-Hill Companies.)

of the pipe. The resistance to flow is analogous to friction but without a dependence on normal stress.

The hydraulic conductivity of soils can vary by orders of magnitude, so consolidation of sand can take place almost immediately and consolidation of clays can continue for decades. Pores between soil particles are interconnected, the shapes are extremely irregular, and areas can be further restricted by water immobilized on clay mineral surfaces. Terzaghi therefore proposed a *coefficient of consolidation, C_v*. Through a consideration of change in flow rate, $\partial u/\partial t$, it can be defined on the basis of the distance to a boundary where the flow rate is zero:

$$\frac{\partial u}{\partial t} = C_v \frac{\partial u^2}{\partial z^2} \tag{3.4}$$

This equation is derived in most geotechnical engineering textbooks. The rate term at the left represents the decline in pore water pressure with time, and the term at the right relates to changing slope of the curves in Fig. 3.4, which defines a theoretical relationship between flow rate, u, and distance, z.

Terzaghi simplified the time calculation by incorporating a *time factor, T*:

$$t = \frac{H^2}{C_v T} \tag{3.5}$$

where t is time, H is the maximum distance to a drainage face, T is a time factor, and C_v is the *coefficient of consolidation.*

Defining a Drainage Distance

In Eq. (3.5) consolidation time depends on the *square* of the drainage distance. With two-way drainage, H is one-half of the consolidating layer thickness. Thus

$$\frac{t_{\text{field}}}{t_{\text{lab}}} = \frac{|H_f|^2}{|H_{\text{lab}}|} \tag{3.6}$$

where subscripts represent field and laboratory maximum drainage distances.

Example calculation of time equivalence: A 3-m (3000-mm)-thick layer of clay between two layers of sand is modeled in a consolidation test with a specimen having an initial thickness of 30 mm. What is the equivalent field time for 24 hours in the lab test?

$$H_f = 3000/2 = 1500 \text{ mm}$$

$$H_{\text{lab}} = 30/2 = 15 \text{ mm}$$

Answer: The time equivalence is $t_f / t_{\text{lab}} = (1500/15)^2 = 10{,}000$, and a day in the laboratory therefore can represent 10,000 days = 27.4 years in the field.

3.7. Pore Water Pressure and Foundation Bearing Capacity

Foundation bearing capacity depends on the resistance of soil to shearing and is discussed in a later chapter. According to the concept of effective stress, pore water pressure subtracts from normal stress along a shear surface:

$$\sigma' = \sigma - u \tag{3.7}$$

where σ' is the effective normal stress, σ is the total normal stress, and u is the pore water pressure. *The critical time for a bearing capacity failure therefore is likely to be during first loading when pore water pressure is highest.*

Field Monitoring

Because pore water pressure can be a critical factor affecting foundation stability, it often is monitored with *piezometers* installed at critical depths and place, such as in the middle of a consolidating layer close to the area of loading. Piezometers are pipes with filters at the lower ends. They in effect are miniature wells, so pore pressure is measured from the height to which water rises in the pipe. This can be monitored electronically or measured with a metal tape making electrical contact with the water.

3.8. Pore Water Pressure Dissipation and Rate of Primary Consolidation

A theoretical distribution of excess pore water pressure in a clay layer is illustrated in Fig. 3.4, where drainage is in two directions, up and down. The curve $U = 50$ percent separates equal areas, and shows a theoretical distribution of excess pore water pressure in

Consolidation Completion, Percent	T
0	0.00
50	0.197
90	0.848
95	1.13
100	~3.0

TABLE 3.2 Values of T for use in Eq. (3.8)

a clay layer when consolidation is 50 percent complete. Other percentages based on the areas defined by the area ratios are the basis for T values in Eq. (3.2), and are shown in Table 3.2. A time relationship can be expressed by

$$t = \frac{H^2}{C_v T} \tag{3.8}$$

where t is time and H is the maximum drainage distance, which in many cases will be one-half of the layer thickness, and C_v is a soil *coefficient of consolidation*. A derivation of this equation is presented in most geotechnical engineering textbooks. A common reference value used in geotechnical reports is the time for 90 percent completion, in which case Eq. (3.8) becomes

$$t_{90} = \frac{0.85H^2}{C_v} \tag{3.8a}$$

3.9. Evaluating C_v

The interpretation of consolidation test data was pioneered by Prof. Arthur Casagrande. Figure 3.2 shows the time-dependency of consolidation after a load is applied. A similar graph can be prepared for each pressure applied in a consolidation test, in order to relate the amount of consolidation to soil pressures anticipated to occur at various depths under a foundation. The distribution of foundation pressures is discussed in Chap. 5.

The soil coefficient of consolidation, C_v, can be evaluated from t_{50}, which is the time required for 50 percent primary consolidation in a consolidation test. The value determined for t_{50} is substituted into a rearranged Eq. (3.8):

$$C_v = \frac{0.197H^2}{t_{50}} \tag{3.8b}$$

where H is the maximum drainage distance, or one-half of the specimen thickness in a consolidation test.

Example calculation of a coefficient of consolidation: In Fig. 3.3, t_{50} = 6 minutes. With an initial specimen thickness H = 17.13 mm,

$$C_v = 0.197 \times (17.13/2)^2 \div 6 = 9.6 \text{ mm}^2/\text{min}.$$

3.10. A Reference Time for 90 Percent Primary Consolidation

After C_v has been determined it can be used with theoretical T values in Table 3.2 to estimate other consolidation times.

Example of an equivalent field time: With the value of C_v from the previous example, estimate the time for 90 percent primary consolidation of a 1-m-thick layer of consolidating clay between two layers of sand.

Answer: One-half of the consolidating layer thickness is H = 0.5 m. From Eq. 3.8(a) we have

$$\begin{aligned} t_{90} &= \frac{0.85(0.5\,\text{m})^2}{9.6\,\text{mm}^2/\text{min}} \\ &= \frac{0.85(500\,\text{mm})^2}{9.6\,\text{mm}^2} \\ &= 22{,}135\,\text{minutes} \\ &= 15.4\text{ days} \end{aligned} \tag{3.8c}$$

Question: How long would it take if there is only one drainage direction?

Answer: $(1\text{ m})^2 / (0.5\text{ m})^2 = 4$, which gives $4 \times 15.4 = 61$ days.

3.11. It's Not Over Until It's Over: Secondary Consolidation

The seepage rate mechanism that controls primary consolidation obviously does not hold for secondary consolidation that occurs after all excess pore water pressure has been dissipated. Attempts to link secondary with primary consolidation are clouded because of the two different control mechanisms.

If consolidation tests indicate that because secondary consolidation will be small, its effect may be included a soil report with a qualifying statement such as "...over a long period of time, a small amount of long-term secondary consolidation and settlement can occur in addition to the amount of primary consolidation settlement calculated in this report." The final amount can readily be calculated using a *first-order rate equation.*

3.12. First-Order Rate Equations

A first-order rate relationship can be defined by a rate that is declining and is proportional to the departure from an end value. The relationship was discovered in the 1860s when the rate of chemical reactions was discovered to slow down as the reactions run out of a reactant.

First-order rate equations also have been applied to physical processes that are slowing down as they approach a hypothetical ultimate value. For example, there appears to be a limiting value for life expectancy. Also to how fast a person can run or how high he or she can jump in a particular field of gravity. The equation can be tested by measuring the speed of a vehicle that is allowed to coast to a stop on a level surface: The speed will diminish rapidly at first, then decline more slowly as wind and other resistances decrease, and roll toward an asymptotic stop. The method is based on Newton's Second Law. It can be used where other methods fail, as for measuring the resistance of sea ice rammed into by an icebreaker that comes to a stop (Handy 2002, 2013).

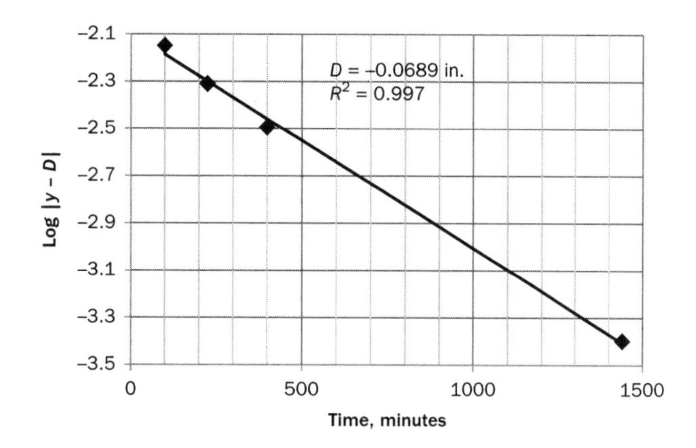

FIGURE 3.5 FORE solution for final value of settlement that includes secondary consolidation of data in Fig. 3.3.

A similar relationship can be shown for primary consolidation data and can be attributed to an asymptotic decrease in the resisting pore water pressure. For secondary consolidation the control may be a gradual increase in bonding between particles in accordance with the adhesion theory of friction. A FORE equation applied to either type of consolidation of soils can be expressed as

$$\log|y - D| = kt + C \tag{3.9}$$

where y is the rate measured at time t, D is a projected final value, and k and C are constants.

In the equation there are two measured values, y and t, and three degrees of freedom, D, k, and C, which means that additional information is required to obtain a solution. The added information is because the equation is linear. It therefore can be solved by substituting trial values for D and testing for linearity. This is conveniently done with a spreadsheet graph, as it can instantly calculate statistical R^2 values with trial values of D, which should approach 1.00 for a perfect linear fit. A computer spreadsheet solution and graph of Eq. (3.9) can almost instantly calculate R^2 values to enable selection of the best-fit D.

Example of FORE applied to consolidation test data: FORE can be applied to both primary and secondary consolidation. Figure 3.5 shows a solution for secondary consolidation of data in Fig. 3.3 with D that gives the highest R^2 value. The projected final settlement is $D = -0.0689$ in. (1.75 mm), which compared with 33 mm cited in the example following Sec. 3.3 is 5 percent of the total projected settlement.

3.13. Field Time for Secondary Consolidation

Although primary consolidation depends on declining pore water pressure that is separating soil grains, secondary consolidation can only start after the grains come into contact. The rate depends on the rate of deformation of the grains themselves, which

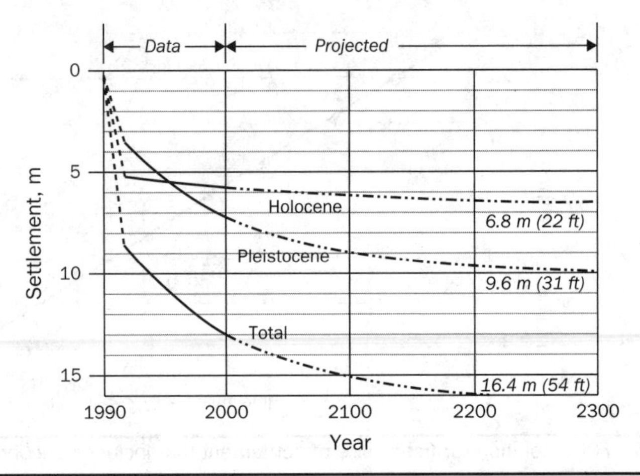

FIGURE 3.6 Settlement amounts of an artificial island in Japan, as measured and as projected into the future by FORE. (Image source: Adapted from Handy (2002). Used with permission from the American Society of Civil Engineers.)

is difficult to predict. As the rate depends on contact stresses, it can be influenced by a groundwater table so secondary consolidation may only proceed when the water table is low. The rate also should depend on pressures created by foundations that are discussed later in this chapter. That makes times for secondary consolidation difficult to predict, and the most important calculation may be the amount that eventually can be expected to occur. Select examples have indicated that soft thick clay layers can continue to consolidate for years. The rate also will be intermittent as it relates to effective stress that can be depending on the elevation of a groundwater table.

Field Data

The Kansaii International Airport at Kobe, Japan, was constructed on an artificial island on soft ocean bottom sediment. Vertical sand drains were installed in the upper layer of soft clay to hasten primary consolidation but their depth was limited. Figure 3.6 shows the most severe settlement that was measured for a period of 10 years and extrapolated to 310 years using FORE. Settlement of the upper, "Holocene" layer penetrated by sand drains occurred rapidly and then suddenly slowed and entered secondary consolidation. The FORE projection indicates that eventually part of the island will require additional fill to remain above sea level. A second island has since been constructed. (Islands in politically disputed areas of the South China Sea are on coral reefs.)

3.14. Defining a Preconsolidation Pressure

Settlement can be minimized if a soil preconsolidation pressure is not exceeded. The usual causes are removal of overburden pressure by geological erosion, melting of continental glaciers, lowering of a groundwater table, and construction activities including soil compaction, removal of overburden, and supporting heavy construction equipment. Most soils to some extent are overconsolidated.

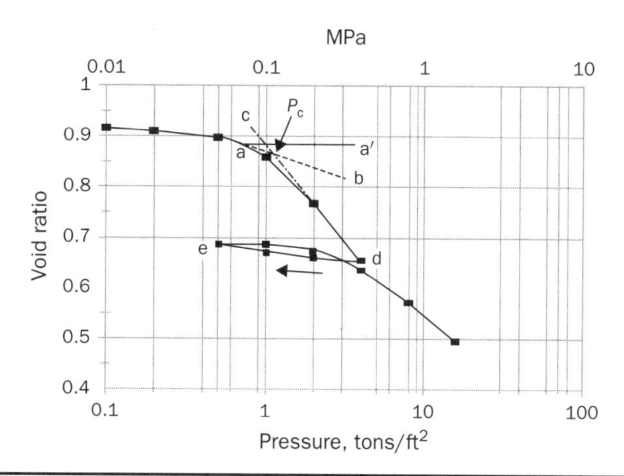

Figure 3.7 The Casagrande method for estimating preconsolidation pressure. (Image source: *Geotechnical Engineering: Soil and Foundations Principles and Practice* by Richard L. Handy and Merlin G. Spangler. © 2007, McGraw-Hill Companies.)

Casagrande Method

A preconsolidation pressure can be created during a laboratory consolidation test with an unload–reload cycle; see ed in Fig. 3.7. A simple graphical method was suggested by Arthur Casagrande for estimating preconsolidation pressure:

1. Select a point of maximum curvature and draw a horizontal line aa′.
2. Extend line cd back from the linear sloping part of the graph.
3. Draw a line ab to bisect the angle.
4. The intersection of ab with cb is the estimated preconsolidation pressure.

Correcting for Sample Disturbance

Schmertmann (1955) performed consolidation tests on samples with varying amounts of disturbance and developed a procedure for adjusting a consolidation test curve using an unload–reload cycle:

1. Calculate the initial void ratio e_0 from the specimen density and moisture content prior to loading.
2. As shown in Fig. 3.8, draw a horizontal line from e_0 to E, which is the calculated effective stress from the overburden pressure prior to loading. Line e_0E is assumed to represent the virgin compression curve.
3. Plot the Casagrande P_c on line e_0E. Extend a line back from the linear sloping part of the graph to intersect the horizontal line at a point designated as E'.
4. Draw a line starting at E' parallel to the line established by the unloading test. Point m is the minimum P_c.

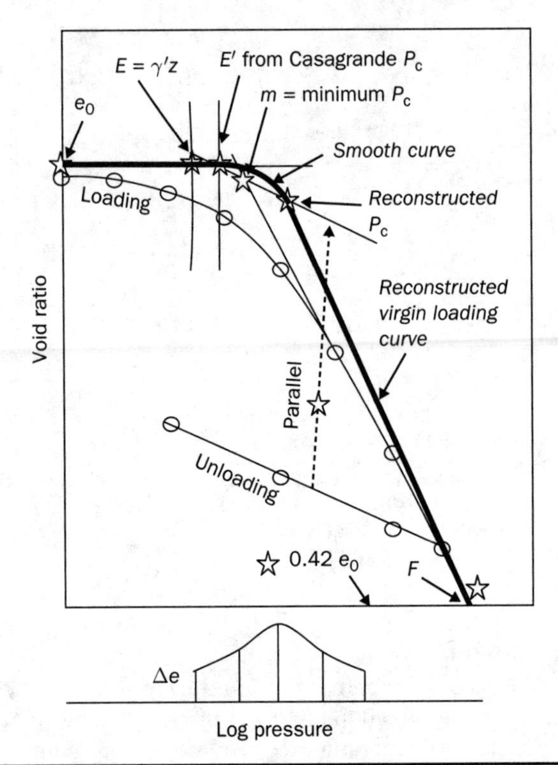

FIGURE 3.8 The Schmertmann empirical correction to consolidation curve to account for sample disturbance. (Image source: *Geotechnical Engineering: Soil and Foundations Principles and Practice* by Richard L. Handy and Merlin G. Spangler. © 2007, McGraw-Hill Companies.)

5. Draw a line from m to point F, which is on the test curve where the void ratio is $0.42 \times e_0$. (It is found experimentally that there is little change at pressures in excess of this value.)

6. P_c should be at a point of maximum pressure difference between the actual test curve and the reconstructed curve, shown by the diagram at the bottom of the figure. If it is not, an adjustment is made and the process is repeated.

Use and Misuse of OCR

The overconsolidation ratio, commonly designated OCR, is defined as the preconsolidation pressure divided by the overburden pressure on an effective stress basis. Sometimes a *soil* is described as having a particular OCR, which is incorrect because it is the *soil sample* that has a particular OCR. The reason is obvious: The overburden pressure is the denominator and equals the sample depth times the unit weight of the overlying soil on an effective stress basis, so if the depth is zero the OCR of every overconsolidated soil is the same; it is infinite.

3.15. Lambe's Stress Path Approach to Settlement

Horizontal in situ soil stress is not reinstated in a consolidometer test; so Prof. T. W. Lambe at M.I.T. devised a "stress path" procedure to try and duplicate a soil stress

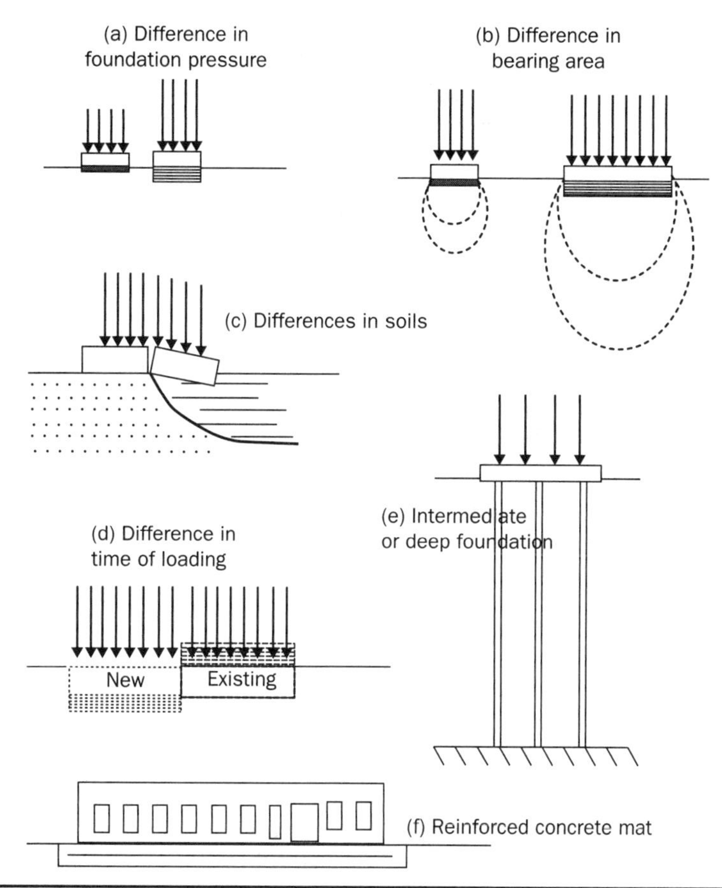

FIGURE 3.9 Common causes of differential settlement and some preventative measures. (Image source: *Geotechnical Engineering: Soil and Foundations Principles and Practice* by Richard L. Handy and Merlin G. Spangler. © 2007, McGraw-Hill Companies.)

history with a laboratory triaxial shear test. Vertical stress is calculated from elastic theory. The method is most useful for normally consolidated soils where expected settlement is at a maximum and lateral stress is predictable from the Jáky equation:

$$K_{onc} = 1 - \sin\phi \tag{3.10}$$

where ϕ is the angle of internal friction. A simplified derivation is in Handy and Spangler (2007, p. 554).

3.16. Differential Settlement

Some common causes of differential settlement are shown in Fig. 3.9*a–d*, and common preventative measures are shown in Fig. 3.9*e,f*.

Problems with Building Additions

Foundation pressures can be identical, and the addition almost inevitably settles more than the original building unless it is supported on a deep foundation such as piles. The difficulty of course is because the existing building already has settled. If the accepted design value is 1 in. (25 mm), it can be accommodated by vertical joints between connecting walls. Joints in exterior walls should be sealed with suitable mastic. Floor connections with short ramps can prevent tripping.

3.17. The Other Shoe

Defining soil behavior under load is only half of the problem. The other half involves defining the distribution of foundation stress in soil; this is discussed in Chap. 5.

Problems

1. How may one distinguish between primary and secondary consolidation?
2. A building is designed with uniform pressures on all wall and column footings but there is damaging differential settlement. Explain.
3. What steps should be taken to avoid or prevent differential settlement?
4. A lawsuit revolves around whether damages to a house were caused by differential settlement or by expansive clay, neither being covered by homeowner insurance. Differential settlement would establish liability and expansive clay usually is regarded as a seller responsibility. Suggest an appropriate investigation.
5. Can expansive clay consolidate?
6. The void ratio of a soil is 0.798 at a pressure of 95.8 kPa and the compression index is 0.063. Draw a void ratio–pressure diagram for this soil on both semilogarithmic scale and natural scales.
7. A clayey soil has a liquid limit of 28. Estimate the compression index and explain its meaning, limitations, and how it can be used to predict settlement.
8. Is there any limitation to the use of a compression index with an overconsolidated soil?
9. A compressible soil layer is 8.5 m (28 ft) thick and its initial void ratio is 1.046. Tests and computations indicate that the final void ratio after construction of a building will be 0.981. What will be the settlement of the building over a long period of time?
10. A square column footing, founded on sand at a depth of 1.2 m (4 ft) below ground level, will carry a load of 890 kPa (100 tons). The average standard penetration value of the sand is 20 blows/0.3 m (ft). Make a preliminary determination of the size of footing to the nearest 1 ft (0.3 m) that will be required to limit settlement to 1 in. (25 mm), (a) if the watertable is 3 m (10 ft) below ground level, (b) if the watertable is at ground level.
11. Use FORE to project a final value and R^2 for the following data obtained with constant time intervals: 88, 44, 22, 11.

12. Both theoretical and empirical approaches are used to address problems in this chapter, so which is better? Why is there such a shortage of hypotheses and theories and an abundance of opinions? Is a theory just a theory?
13. How did Terzaghi's background in mechanical engineering influence foundation engineering?
14. How does overconsolidation affect horizontal in situ stress? Is there a formula?

References

Bowden, F. O., and Tabor, D., *The Friction and Lubrication of Solids,* Oxford University Press, New York, NY, 1950.

Handy, R. L., "First-Order Rate Equations in Geotechnical Engineering," *ASCE JGGE* 128(GT5):416–425, 2002.*

Handy, R. L., *FORE and the Future of Practically Everything,* Moonshine Cove, Ashville, SC, 2013.

Handy, R. L., and Spangler, M. G., *Geotechnical Engineering,* 5th ed., McGraw-Hill, New York, 2007.

Schmertmann, J. H., "The Undisturbed Consolidation Behavior of Clay," *ASCE Transactions,* 120:1201–1233, 1955. Reprinted in *From Research to Practice in Geotechnical Engineering.* ASCE Geotechnical Special Publication No. 180:1.

Terzaghi, K., *Erdebaumechanik,* Franz Deuticke, Liepzig und Wien, 1925.

Further Reading

Casagrande, A., "Karl Terzaghi, His Life and Achievements," In *From Theory to Practice in Soil Mechanics,* L. Bjerrum et al., eds. John Wiley & Sons, New York, 1960.

Schmertmann, J. H., Hartman, J. P., and Brown, P. R., "Improved Strain Influence Diagrams," *ASCE Journal of Graphic Engineering and Design,* 104(GT8):1131–1135, 1978.

Terzaghi, K., *Theoretical Soil Mechanics,* John Wiley & Sons, New York, 1943.

* A critical article reversed the X and Y axes so there were no solutions, and the closure stated that subtracting a constant before taking the logarithm is only a "matter of semantics," which obviously is not correct.

Arthur Casagrande (1902–1981). A professor at Harvard University and pioneer in geotechnical engineering.

CHAPTER 4

Soils Behaving Badly

...And What to Do About It

4.1. Expansive Clays

Expansive clays expand when wet and shrink when dry, and both can create problems for floors and foundations. Problems may develop slowly and over a longer period of time, as when a layer of concrete cuts off evaporation so the underlying clay accumulates water and expands. Uplift pressures can be sufficient to lift floors and foundations and do so unevenly. Walls develop diagonal cracks, and rectangular openings for door and window frames gradually become distorted into parallelograms. The severity of the damage depends on the amount and kind of expansive clay mineral, its chemistry, time, and the availability of water.

Expansive Clay in a Consolidation Test

Figure 4.1 shows consolidation test data for a specimen of expansive clay that was 1.25 in. (32 mm) thick when it was placed in a consolidometer at its natural moisture content. When water was introduced, the soil expanded until the specimen was 1.425 in. (36 mm) thick, a volumetric expansion of about 12 percent. A pressure of 3.3 Tsf (0.33 MPa) was required to compress the specimen back to its initial thickness.

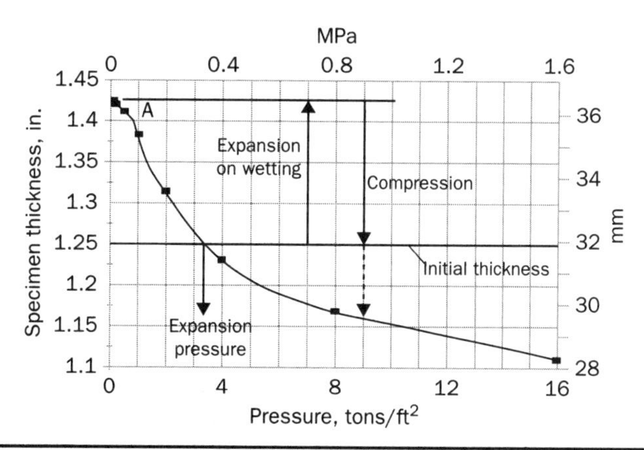

FIGURE 4.1 A clay soil behaving badly: When water was introduced in a consolidation test, the soil expanded, and a pressure of about 0.34 MPa (3.4 Tsf) was required to return it to its initial thickness.

4.2. Two Classes of Expansive Clays

The annual cost of repairs to damages from expansive clays is measured in *billions* of dollars annually in the United States alone. Expansive clays can be classified as *Type G* if they are deposits having a geological origin, and *Type P* for an origin that involves in-place weathering and destruction of non-clay minerals. "Pedology" literally means soil science. The main emphasis is on agricultural but it cannot safely be overlooked in engineering.

Type G Clays

Expansive clays having a geological origin are soft rocks that can be tens of meters thick and millions of years old. They usually are identified as *claystone,* or as *shale* if thinly bedded. Layering is a result of prolonged burial under large amounts of overburden that has been removed by erosion, as the weight has caused clay mineral particles to lay flat. Geologically old shale often is interbedded with coal and has been chemically modified so it is less likely to contain expansive clay minerals.

Type P Clays

Type P expansive clays are developed at a ground surface by weathering, and typically are a meter or two thick. If buried under younger deposits they become *paleosols,* or ancient soils.

A typical weathered *soil profile* consists of a topsoil layer, or *A horizon,* on top of a clayey subsoil *B horizon* layer. Topsoil contains organic matter that gives poor engineering qualities so it normally is stripped prior to construction and spread later to support lawns. The subsoil *B horizon* typically is a meter or two thick and contains a concentration of clay produced by weathering and washed down from the A. It is the subsoil B horizon that is most likely to have a concentration of expansive clay minerals.

How a Layer of Expansive Clay Can Cause Trouble

Most construction takes place in a dry season when clay is not fully expanded. Covering with an impermeable layer such as concrete prevents evaporation, so the clay expands and lifts the concrete. Lifting is uneven because evaporation still can occur around edges. A clue to expansive clay is concrete aprons that extend slab edges. Another clue is above-ground burials in cemeteries.

As topsoil and subsoil layers follow contours of hills, floors and foundations that are horizontal can cut across an expansive clay layer with predictable consequences.

Nature's Color Coding

Soils developed by intense weathering in tropical and subtropical climates usually are not expansive. A clue is a rusty-red color created as weathering destroys expansive clay minerals and releases their iron atoms to make iron oxides coatings on soil grains.

The clay mineral in weathered tropical soils usually is *kaolinite,* which in a pure form is used to make porcelain. Prolonged intense weathering can break down the kaolinite and create *bauxite,* aluminum ore.

Not all soil in tropical areas is highly weathered, as it can be carried in and deposited by rivers as floodplains and rice paddies. Expansive clay minerals still can dominate those areas and add to agricultural productivity while subtracting from engineering uses.

4.3. Sorting Out Floodplain Clays

Many major cities were founded where there is access to transportation by water so they are on floodplains. Floodplains by definition flood, repeatedly, to build up a variety of soil deposits that are more interesting than amusing.

What Makes River Floodplains Wide

Floodplains are widened by rivers cutting away hillsides along the edges, but even more significant is a rise in sea level when the last continental glacier melted. This is discussed in Chap. 1.

While sea level was lower, major rivers entering the seas entrenched and cut wide canyons. Rivers carrying glacial outwash from the final meltdown deposited sand and gravel as sea level came up, because of a natural tendency of water to not run uphill. Sand and gravel then proceeded to be covered later with an assortment of later floodplain deposits including clay. It no coincidence that major bridges on major floodplains are supported on deep foundations extend down through the clay.

Braided Rivers

Rivers where there is a shortage of water become choked with sand and gravel, so channels continuously shift around leave almond-shaped sand bars that create a braided appearance. Braided rivers have a high downhill gradient and flow fast to keep gravel and coarser materials moving, so they can aggressively attack a river bank that is not protected. A braided river carrying glacial outwash in Alaska is shown in Fig. 1.4. Rivers carrying glacial outwash were (and still are) sources for wind-deposited loess.

Meandering Rivers

Rivers with no shortage of water can reduce the velocity and downhill gradient by meander around with no apparent sense of direction, but they actually *do* have a sense of direction. Meanders keep pushing to the right (looking downstream), so floodplains tend to be narrow on the right side of a river and narrower on the left. This is in response to the *Coriolis effect*, first described in 1835 and named after a French scientist and engineer, G. G. Coriolis (1793–1843). He examined a matter suggested about 200 years earlier by Isaac Newton and known as his First Law, which is fairly simple, and states that unless acted upon a moving object will go in a straight line. That means that on a revolving earth the surface may move out from under it. The same principle applies to rotations of winds in hurricanes and of water in a flushed toilet. It also is consistent with traffic driving on the right. A meandering river is shown in Fig. 1.3.

A Shift from Braided to Meandering

After the final glacial meltdown, rivers carrying outwash responded to a reduced sediment load by changing from braided to meandering that extended their length and lowered the velocity created by a downhill gradient. That left patches of sand and gravel in high terraces flanking the floodplains that are sources for sand and gravel.

4.4. Floodplain Soils of Meandering Rivers

Floodplain soils are among the most diverse and challenging encountered in foundation engineering, and the engineer who ignores this is likely to have a challenging career. Meandering rivers can deposit a host of soils ranging from sand to silt and clay that often is expansive. The identification and nature of the soils depends on the manner of deposition.

Oxbow Lake Clay

Least desirable is very soft clay deposited in an oxbow lake. It may have a hard crust but it can be deceiving. An oxbow lake is an abandoned river meander, created when the downstream migration of a meander is held back so the next loop back catches up. The river immediately takes a shortcut across a sand bar, usually during high water when the river channel is eroded at its deepest. The increase in gradient at the *cutoff* and the fact that it is through sand cause it to immediately erode and entrench, so once formed there is no going back. Ends of the isolated meander connected to the river immediately plug with sand so the abandoned meander becomes isolated as an oxbow lake.

Clay carried into during high water is trapped when the water level goes down, so buoyancy slows settlement and reduces consolidation under its own weight. This is particularly true for expansive clay because it has a low unit weight when fully expanded.

Oxbow lake clays are avoided if at all possible, and they must be recognized before they can be avoided. Because they can be relatively narrow, they can be missed by routine soil borings, particularly if the location carries the inconvenience of a drill truck getting stuck. Many firms now employ ATV's (all-terrain vehicles) for drilling in difficult areas.

Depth and Shape of an Oxbow

A typical oxbow lake is curved so it approximates a half-circle. The clay fill is a cast of the river when it was eroding deepest, and is shallow along the inner bank and deepest along the outer bank that was turning the river. Where an oxbow is obscured under vegetation or difficult to perceive from the ground, it still can be evident on drone or air photos and on photo-based USDA county soil survey maps.

Slack-Water or Backswamp Deposits

Second on the list of undesirables but often unavoidable is floodplain clay that is deposited after each flood. The clay often is expansive and can be expensive. It can reach a considerable thickness from being deposited layer-on-layer. It also can cover and obscure oxbow lake clay. Other less offensive floodplain deposits including sand *point bars* enclosed by meanders and silt *natural levees* deposited on floodplain around the outer rims of meanders are discussed in Chap. 1.

> *Conquering the 100-year flood.* The frequency of 100-year floods is calculated from a statistical analysis of historical data. However, even if a floodplain is protected by dams and levees, it is not unusual for the 100-year flood to repeat every few years. May there be a relationship to ocean temperatures? (A 1° C increase in the temperature can increase water vapor pressure about 6 percent.)

4.5. Deep Tropical Weathering and Expansive Clay

Massive eruptions of basalt lava emerged from fissures in the earth's crust in India, at about the same time that an asteroid rang the earth's bell in Yucatan, so there may be a connection. According to a principle suggested by French chemist Le Chatelier (lay-shot-lee-ay) (1850–1936), a material that is in equilibrium at high-temperatures is likely to be less stable and more easily weathered at low ones.

The Deccan Plateau in India is one of the world's largest continuous areas of expansive clay. Even though the climate is hot, the clay is black, not red, and it is highly expansive. Weathering has been arrested by vertical mixing connected with seasonal shrink-swell cycling. The Columbia Plateau in the northwest corner of the United States also is composed of lava flows, but they are more recent and soil was washed clean by glacial meltwater to leave bare "scablands."

4.6. A Guide to Expansive Clay

Figure 2.6 shows relationships suggested by Chen (1988) based on clay samples placed in a consolidation test apparatus and then wetted with water. The tests were conducted on cut (undisturbed) specimens. The loss of a restraining soil structure through disturbance is likely to make them more expansive.

Crystal Structure in Control

Most minerals do not have fixed chemical formulas, least of all expansive clay minerals. They also have positive ions such as calcium, Ca^{++} or sodium, Na^{+} that occur between layers and control the amount of expansion.

Expansive clays are referred to as *montmorillonite group minerals* or *smectites.* A simplified diagram showing layered crystal structure of mica and expansive clay is shown in Fig. 4.2, where corners of the shaded triangles represent centers of oxygen atoms that form the framework for the crystal structure. The crystal structure has a net negative electrical charge that is balanced by positive ions between the layers and which tend to hold the layers together. Most efficient in this respect is potassium, K^{+}, which is a large ion and makes a close fit that prevents expansion. The mineral is mica.

The most common interlayer ion, or cation, in naturally occurring expansive clays is calcium, Ca^{++} which can limit the amount of expansion. Air-dry, the clay has 2 layers of water, and saturated it has 3, as shown in Fig. 4.2. The affinity for water can create sufficient uplift pressure to lift floors and foundations. The expansion state depends on the relative humidity, discussed later in this chapter.

A forced exchange of sodium, Na^{+} for calcium, Ca^{++}, cuts expansion loose and is used to disperse clay for a particle size analysis based on settlement rate in water. *Bentonite* is a naturally occurring sodium expansive clay created by weathering of volcanic ash. The clay is a component of drilling *mud* that is pumped down a drill pipe to bring up drill cuttings along the outside. If drilling is interrupted, the mud "sets" and prevents cuttings in transit from settling and jamming the drill.

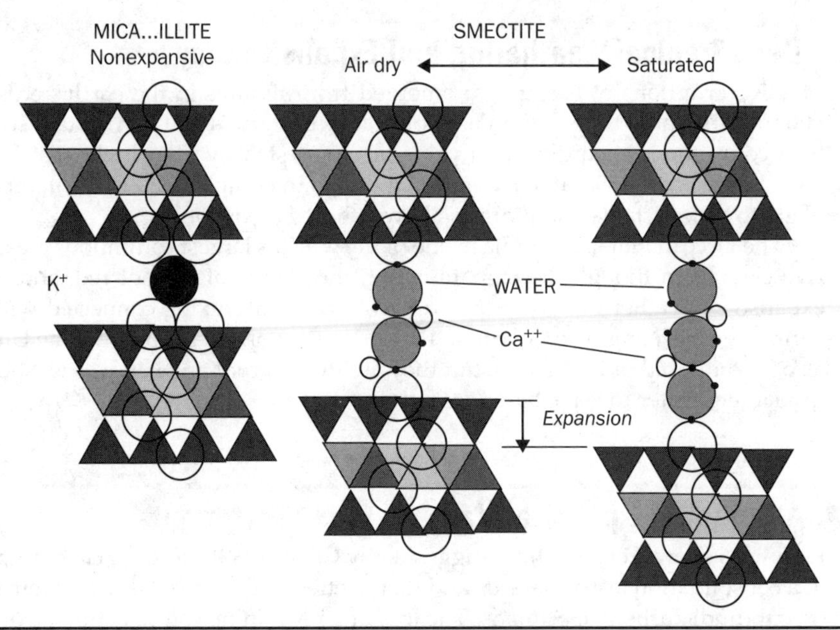

FIGURE 4.2 A simplified illustration of clay mineral expansion: Mica has a layered structure but is not a clay mineral, as its sheets are held together by positively charged potassium ions. Weathering can substitute calcium ions that allow limited expansion. Black dots represent hydrogen atoms on water molecules.

4.7. Field Evidence for Expansive Clay

Expansive clay exposed on dry lakebeds can show an obvious array of vertical tension cracks, and to preserve energy the cracks intersect in a honeycomb pattern. A similar crack pattern exists in Arctic *permafrost,* created by thermal volume changes in frozen soil.

Tension cracking is less obvious when concealed under vegetation. Clay subjected to repeated shrink–swell cycles can develop a "subangular blocky" soil structure. It occurs in a subsoil "B horizon" that typically is 1-2 m thick and follows ground surface contours. The damaged clay has poor engineering qualities because individual blocks are preserved with thin films of clay, or "clay skins," that prevent bonding. A subangular blocky structure can cause a clay soil to be misidentified as sand in cone penetration tests.

More Bad Karma

Open tension cracks are an open invitation for debris and loose soil to fall in and prevent complete closing when clay between the cracks expands. Repeated cycling causes lateral pressure to build up until it exceeds the unconfined compressive strength of the moist clay so it shears along inclined surfaces that become smeared with clay "slickensides" as shown in Fig. 4.3. Such surfaces can severely affect engineering behavior, and shearing causes vertical mixing so topsoil and subsoil combine into a thick, black, clay layer. The common name is *black cotton soil;* the scientific name is *Vertisol,* for vertical mixing. Compaction in layers with tamping or sheepsfoot rollers can break up the shear planes, but the soil still is expansive.

FIGURE 4.3 Inclined shear surface preserved by "slickensides" created by horizontal expansion pressures in expansive clay. Shearing causes vertical mixing and permanently weakens the clay.

4.8. Managing Expansive Clay

The Chainsaw Method

Damages from expansive clay may be postponed until trees are large enough for roots so suck appreciable amounts of water out of the soil. An option follows a hard line and requires diplomacy, and will do nothing to correct existing damages.

Structural Slabs, Grade Beams, and Piles

Concrete floor slabs on expansive clays can be supported on a waffle-like arrangement of reinforced grade beams. Grade beams supported on piers that extend down through the active layer often are framed and poured on top of crushable cardboard "Verticel," and a substantial air gap must be provided between the floor and the soil. Another procedure is to reinforce and integrate the beams with the floor, as shown in Fig. 4.4.

Stripping off the Active Layer

The "active layer" of expansive clay undergoes seasonal shrink–swell cycles, so it might be assumed that removing the active layer might solve the expansive clay problem, but as shown in Fig. 4.5, stripping it off but only sends the shrink-swell activity deeper where it remains destructive.

FIGURE 4.4 Construction of a structural floor with integrated grade beams on expansive clay. After Briaud et al. (2016). Available online: ceprofs.tamu.edu/briaud/Iran 2010.pdf

FIGURE 4.5 Excavation for the Malaprabha Left Bank Canal in expansive clay in India relocated the active clay layer downward, and the canal failed before it was opened. (Inset) The entire canal was reconstructed using Katti's replacement method and has been in service for over 50 years. (Image source: Author photo; inset image courtesy of the Central Board of Irrigation and Power, New Delhi.)

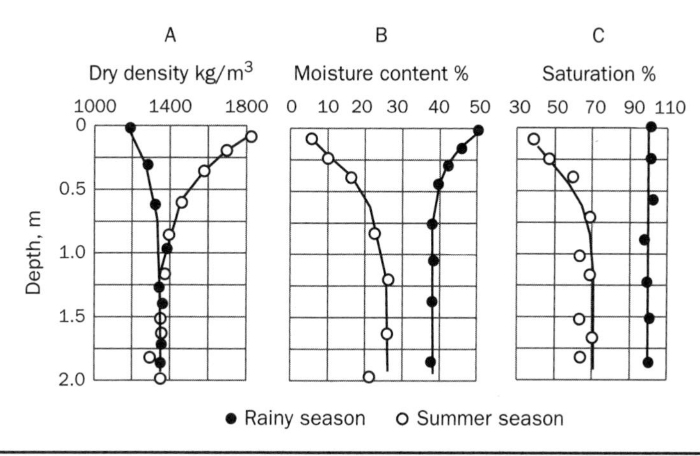

FIGURE 4.6 Moisture contents and densities below a depth of about 1 m provided an important clue that led to a removal-and-replacement method for control of expansive clay. (Image source: After Katti et al. 1969.)

Observations of Strange Field Behavior

Figure 4.6 shows an apparent anomaly that was discovered to be consistent in naturally occurring expansive clays by Prof. R. K. Katti at IIT (Indian Institute of Technology), Mumbai, India. Below a depth of about 1 m, moisture contents change seasonally but do not cause corresponding changes in the soil, and no differences were found in grain size distribution, plasticity or clay mineralogy.

4.9. The Replacement Method

After the canal shown in Fig. 4.5 failed, a test section was discovered that *did not fail.* The section was patterned after the observations shown in Fig. 4.6, but with the upper meter of expansive clay replaced with compacted, non-expansive clay. The canal was rebuilt using that method. It has been in service since 1977, and the irrigation system was instrumental in changing India from being a food importer into a food exporter. The Indira Gandhi canal has since been constructed using the same method. It should be a point of pride for geotechnical engineering.

The process has since been adopted world-wide and called the "replacement method," even though more than replacement is involved. As non-expansive clay usually is not locally available, it is manufactured on-site by adding and mixing in a small percentage of hydrated lime. Other cover materials such as sand are not effective, and concrete obviously is not effective.

How Does It Work?

The method by which stabilization is achieved has remained enigmatic. Katti et al. (1969) suggested that lateral restraint from friction along the contact between layers may play a role but it should rapidly dissipate with vertical distance. Houston et al. (2016) applied *unsaturated soil mechanics* that emphasizes the role of capillary water, but the influence does not stop at 1m depth.

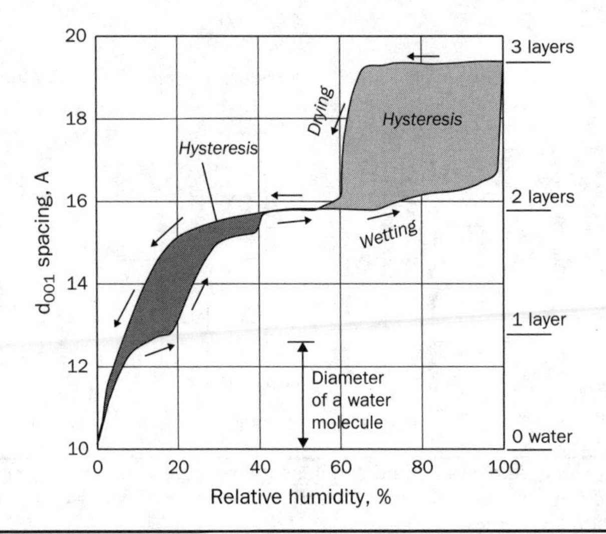

FIGURE 4.7 X-ray diffraction measurements of interlayer spacings in calcium montmorillonite clay as a function of the vapor pressure of water expressed as relative humidity. The lag (hysteresis) affecting the initial removal of interlayer water may explain the "replacement" layer method for controlling expansive clays. (Redrawn from Senich et al., 1965.) *1 Ångstrom = 1Å = 10^{-10} m = 0.1 nm.*

New Rule for Control of Expansive Clay

In unrelated research, samples of pure expansive clay minerals were examined using x-ray diffraction, with the clay placed in a vacuum chamber that could be pumped down until all interlayer water was removed. The dry spacing for calcium smectite (montmorillonite), is shown at the left in Fig. 4.7, and as is the same as that of mica, which has a similar layered structure but is not expansive.

Water vapor then was incrementally introduced back into the system, and caused the mineral to expand, shown by arrows pointing to the right in the figure. Expansion was rapid because there were no competing air molecules. After 1 day the procedure was repeated until the clay was fully expanded, shown at the right in the figure. The process then was reversed and the system pumped back down, shown by arrows pointing to the left in the figure.*

Clues to Between-Layer Stacking of Water Molecules

Expansion paused at interlayer spacings corresponding to about 1 and 2 stacked diameters of water molecules, and stopped at 3 layers. As the system was being pumped down, a lag or *hysteresis* retained interlayer water so about 98 percent relative humidity was required to achieve full expansion. A similar lag affected the removal of interlayer

*This research was conducted by Prof. T. Demirel and Dr. Donald Senich at the Iowa State University Engineering Research Institute, sponsored by the National Science Foundation.

water, as the humidity had to be reduced to about 65 percent to start removal. Therefore if the relative humidity of air entering the system exceeds that percentage, the expanded clay mineral should remain expanded.

Hypothesis

May the function of a cover layer of non-expansive clay only be to process air entering the system so its relative humidity exceeds 65 percent?

a. Measurements of the relative humidity of air in moist soil show that it usually exceeds 98 percent.

b. Soil that is compacted at its optimum moisture content contains air voids so it should be slightly permeable to air.

c. Air that is not fully saturated, as is likely to occur during hot weather, may remove capillary water *without affecting interlayer water and volume change.*

d. A sand surface layer may not retain enough water to be effective, and a surface layer of concrete will only trap water in the soil.

Why Does Clay Expansion Stop at 3 Layers?

Measurements of x-ray diffraction intensities indicate an ice-like structure of water when a calcium montmorillonite (smectite) clay is fully expanded (Senich et al. 1967). The more open arrangement of water molecules also is suggested by the small increase in interlayer spacing starting at about 70 percent relative humidity. Sometimes basic research can provide answers before there are questions.

What's in a Name?

The name "Replacement Method" gives no indication that more than replacement is involved, and other names such as "CNS" for Cohesive Nonexpansive Clay, or "NEC" for Non-Expansive Clay, may be preferable.

4.10. Chemical Stabilization of Expansive Clay with Lime

Non-expansive clay can readily be manufactured from expansive clay by mixing in a few percent hydrated lime, Ca(OH)2. Calcium chloride, CaCl, is not effective, as (OH)- ions are required to pull H+ ions from within the clay structure and make water. That causes the clay particles to become more negative so they are linked by Ca++ ions from the lime. The change is almost immediate, as indicated by a sudden increase in the soil plastic limit so it exceeds the soil moisture content. The change is sufficient to change the engineering soil classification.

Hydrated lime is commonly mixed with clay soil that is compacted for road bases and subbases. Fly ash from burning coal that contains limestone also contains lime and is used as a soil stabilizer. Quicklime introduced into boreholes can permanently stabilize landslides in expansive clay, Fig. 4.8. *A hands-on field test for soil-lime reactivity* is discussed in Section 2.10 of this book.

Figure 4.8 This is the first of many landslides that have been stopped and permanently stabilized with drilled quicklime. The boring spacing is about 10 times the boring diameter, and only the active "pushing" part of a landslide requires treatment. Water can be geysered out of a boring with a scoopful of "hot" pebble quicklime, then the boring is filled with lime except near the top, where it is plugged with soil to contain the expanding lime. As the lime hydrates it dries the soil, expands and injects into radial shrinkage cracks that carry it out into the surrounding soil (Williams and Handy 1967).

4.11. *Collapsible* Soils

Loess, which is from German for loose, is wind-deposited silt. Most loess is deposited from major rivers that were braided streams from carrying glacial meltwater. Major loess deposits in the United States constitute upland hills bordering the Mississippi and Missouri Rivers.

Loess close to a source, Fig. 1.5, usually has such a low density and clay content that when saturated with water it will collapse under its own weight, but structures are built on this soil when there is no alternative, especially where the thickness makes deep foundations impractical.

An obvious requirement is that ponding of surface water, leaky pipes and sewers, open faucets, roadside gutters, and uncontrolled lawn watering must be avoided.

The most reliable method for assessing collapsibility is by loading a soil specimen to its overburden pressure in a consolidometer and introducing water. Collapsible loess usually contains less than about 10 percent clay, classifies as *A-4* or *ML,* and usually has a dry unit weight that is less than 90 pounds per cubic foot (14 kN/m^3). Immediately after deposition tests of loess currently being deposited in Alaska give unit weight can be as low as 65 pcf (10 kN/m^3).

Delayed Collapse

Sometime loess collapse can continue for years or even decades, usually in an area where land developers have leveled terrain by cutting down high areas and filling in low ones, and *it is the fill areas that settles.* Even a properly compacted fill settles while the nearby collapsible soil remains firm and resolute under a foundation load.

A clue to the apparent anomaly can come from mottled gray and brown soil colors below the groundwater table. Gray is as expected, but brown means that at one time the soil was exposed to air above the groundwater table. The sequence was: (1) The basal loess was saturated, gray, and normally consolidated. (2) The added load on top caused the basal loess to consolidate. (3) Water that was squeezed out penetrated upward to saturate and collapse the overlying loess, so the rate of settlement depends on consolidation and the availability of water.

Collapsible Alluvium

And that's not all: *Alluvial fans* are fan-shaped deposits created where streams emerge from uplands and deposit soil. In arid and semiarid areas the soil can be deposited on top of dry soils that can suck water out so the deposits do not consolidate until later when there is the 5-year downpour of rain.

4.12. Regional Changes in Properties of Wind-Deposited Soils

Sand dunes occur close to sources of sand, as grains bounce along on a ground surface. After a dune starts to form, sand grains falling on the lee side lay at the *angle of repose,* which is approximates the angle of sliding friction, which is about 25° for quartz. A *dune tract* is created when sand from one dune feeds another.

Loess deposits can occur downwind from dune tracts, and silt-size particles do not bounce. They can be trapped by vegetation so the deposit retains a low density and is collapsible when saturated with water. Farther from a source where loess is finer grained and thinner, weathering can convert it into expansive clay. The change is reflected in soil liquid limits, Fig. 4.9. The bulges in the figure outline subsoil *B horizons.*

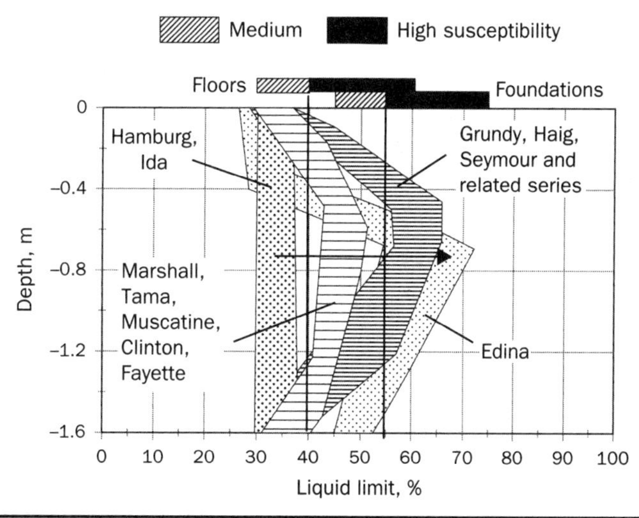

Figure 4.9 The Chen criteria applied to topsoil-subsoil layers in a continuous deposit of loess soil across Iowa. Liquid limit data are from USDA-SCS county soil survey reports. (From Handy (2002), in *Geotechnical News.*)

Figure 4.10 *Accommodation.* As water tends to accumulate more under the center of a structure than the edges, an option for a central bearing wall on expansive clay is to let a supported structure down in small increments. (Image source: Author photo.)

The expansion criteria at the top of the figure are adapted from those of Chen, shown in Fig. 2.6. B-horizon clay that often is expansive *is not included in most engineering soil maps.* Also overlooked can be concentrations of expansive lay on river floodplains. One way to make adjustments for clay expansion is shown in Fig. 4.10.

4.13. Quick Clays!

Quick clays are not like quicksand, where grains are temporarily suspended in a rising current of water. Nor do quick clays contain expansive clay minerals, which would serve as a binder. Instead the clay mineral typically is *illite*, a low-grade mica, and quick clays do not respond to treatment with lime.

Quick clays are deposited in estuaries from salt water that keeps the clay *flocculated,* held together with sodium Na^+ ions, until rains leach them out. As the process is gradual, a quick clay can give the appearance of stability until a sudden vibration sets it off and it slides away as a landslide.

Quick clay is highly *sensitive,* which means that it loses considerable strength when it is disturbed. A simple way to measure sensitivity is with a *vane shear test,* in which a cross-bladed vane such as shown in Fig. 4.11 is pushed into soil and twisted. Ends of the blade often are tapered to minimize the influence from horizontal bedding planes. The vane is pushed vertically into soil to a desired test depth, and torque is applied and increased until the vane rotates. The maximum torque is recorded, and the vane is rotated for several revolutions, and the torque measured again. The ratio of the initial to the remolded strength is the sensitivity. The test was pioneered in the United Kingdom by L. Carlson and A. W. Skempton.

Vane Shear Does Not Just Measure Soil Cohesion

As shown in Fig. 4.11B, pressure from the blades is directed outward so it can mobilize friction. Experiments have been conducted with varying numbers of blades to evaluate soil internal friction but the method is not sufficiently sensitive for use in design.

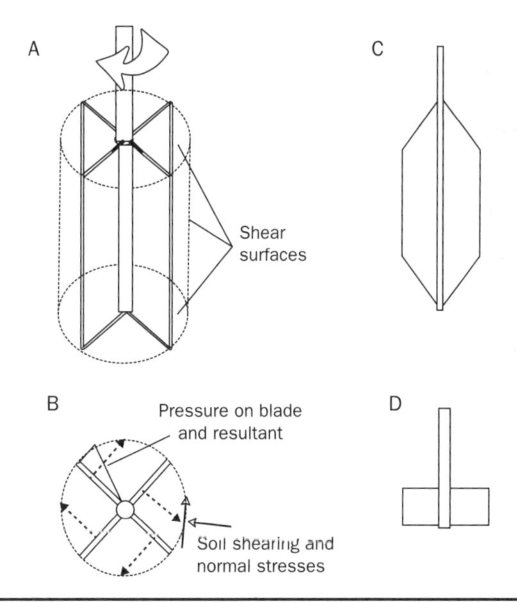

Figure 4.11 (A) Principle of the vane shear test. (B) Outward pressure from rotation of the vane mobilizes soil internal friction as well as cohesion. (C) Design to eliminate the influence from horizontal stratification of the soil. (D) Experimental design to emphasize that influence. (Image source: *Geotechnical Engineering: Soil and Foundations Principles and Practice* by Richard L. Handy and Merlin G. Spangler. © 2007, McGraw-Hill Companies.)

4.14. Liquefaction!

When an earthquake rattles sand grains so they start to slip into voids that are filled with water, the result is a temporary quick condition until the sand grains settle out. The process is called *liquefaction,* and although the condition is only temporary the consequences can be devastating, as structures suddenly tip or sink into the ground. Most vulnerable is "made ground" created by dumping sand fill into water bordering bays and coastal areas and settlement is inhibited by buoyancy.

The longest and most intense period of ground shaking recorded in the United States occurred in 1812–1813 near New Madrid, in the southeast corner of Missouri. It rang church bells across several states, there was settlement, and parts of the river temporarily flowed backward. The underground fault that lay at the cause is closely monitored, as a repeat could severely affect major cities including Memphis.

Identifying Vulnerable Soils

Pioneering studies by Prof. H. B. Seed and his colleagues at the University of California, Berkeley, correlated liquefaction potential of saturated sand to Standard Penetration Test blow count data and the magnitude of earthquakes. Correlations then were developed to Cone Penetration Test data. They are discussed in Chap. 6.

Earthquakes, Volcanoes, and the "Ring of Fire"

Earthquakes and volcanoes are concentrated along plate boundaries in the earth's crust, discussed in Chap. 1. Earthquakes are attributed to stick-slip. Mountain ranges such as the Rocky Mountains and the Andes stand high where a continental plate is overriding an ocean plate. Sedimentary rock layers can be compressed into a series of sine waves that create lines of mountains when tops erode out.

In geological time even granite mountains can be worn down to nubs, and the *shield area* of eastern Canada has been scraped off by continental glaciers.

Made Earthquakes

Pumping water under pressure deep into the ground can relieve effective stress along existing faults so they slip and cause earthquakes. After this was discovered, pumping was stopped. It started up again to get rid of oil-polluted waste water from *fracking*, and Oklahoma now leads the nation in earthquakes.

4.15. Pretreatment to Prevent Liquefaction

Deep Dynamic Compaction involves repeatedly drop a heavy weight to vibrate and pound sandy soil into submission. As shown in Fig. 4.12, cranes can lift and drop approximate 100-ton (900 kN) weights about 100 ft (30 m) so a result was a pattern of waffle-like

Figure 4.12 Addressing soil liquefaction potential with Deep Dynamic Compaction at the Jackson Lake Dam site, Grand Teton National Park, Wyoming. (Image source: Author photo.)

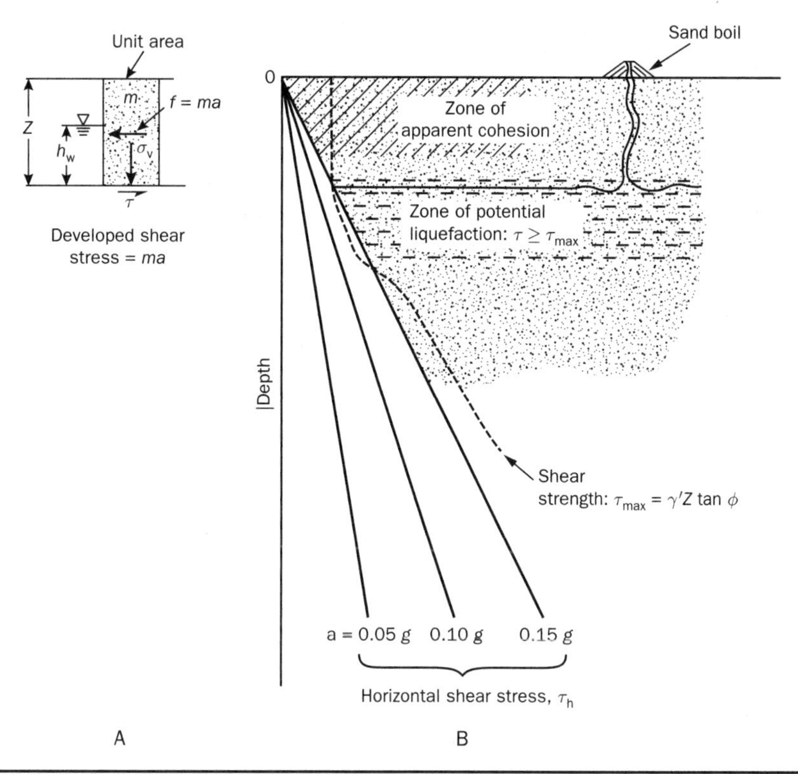

FIGURE 4.13 Increase in earthquake-induced horizontal shearing stress with increasing depth and horizontal acceleration from an earthquake. (Image source: *Geotechnical Engineering: Soil and Foundations Principles and Practice* by Richard L. Handy and Merlin G. Spangler. © 2007, McGraw-Hill Companies.)

depressions that indicate the amount of compaction. Weights may be disconnected for the drop to prevent "backlash" from over-rotation and tangling of cables on the drum.

4.16. Earthquake Dynamics

According to Newton's Second Law, $f = ma$, and the larger the mass m, the larger the force f created by a particular acceleration. As mass increases with depth, the force from a particular acceleration also increases with depth, shown by straight lines in Fig. 4.13. A critical zone for liquefaction therefore can be defined on the basis of soil shearing strength, depth, and vibration acceleration. Liquefaction therefore can occur underground in a zone where vibrations are intensified.

Recurrence Intervals

As sand in a liquefied zone settles out it displaces water that channels upward and deposits small cones of sand the ground surface. This often occurs on floodplains because of an abundance of sand, so the cones then can be covered and preserved under alluvial deposits. Organic matter in the alluvium then can be radiocarbon dated to determine earthquake recurrence intervals related to stick-slip. The intervals unfortunately are not even so recurrence times are statistical estimates that can have a large range.

4.17. Quicksand

Quicksand is *not* a material; it is a *condition*. It is created by upward seepage of water that lifts and separates sand grains so there is no intergranular friction. Despite movie depictions there is no quicksand in a desert where there is no water, and quicksand is a dense liquid so the body floats. Quicksand commonly occurs where a steam with a gravel substratum enters a larger river or lake so there is upward seepage of the water. Quicksand frequently occurs in sandy floodplain soil close to a levee that is holding back a flood. The condition can be held in submission by surrounding the area with sandbags to create a reverse pressure in the water. In the design of dams a quicksand condition near the base can be predicted with a graphical *flow net*.

4.18. Blessed Are the Computers But Will They Really Inherit the Earth?

Computers can map out where failure will initiate. For example, the finite element analysis that is based on elastic behavior will go out of touch when soil starts to shear and lose strength. Color printouts impress clients, but it also is important not to confuse art and accuracy.

Problems

1. In 1980 it was estimated that repair costs to damages from expansive clays were $7 billion in the United States *annually*. Look up inflation rates in U.S. Department of Commerce tables and estimate an equivalent cost today.
2. Expansive clay dominates floodplains and rice paddies in tropical areas even though adjacent upland soil is red-colored and the clay is not expansive. Explain.
3. Identify the following in Fig. 1.5:
   a. Point bar sand deposits enclosed by meanders. What is a clue from agriculture?
   b. Oxbow lake clay. What determines the thickness of the clay?
   c. Natural levee silt deposits on top of river banks around outsides of meanders. The oldest and classic part of New Orleans is on a natural levee of the Mississippi. Why?
   d. A shortcut called a "chute." Why are its meanders smaller than those of the main river?
   e. Why should an oxbow lake clay should be avoided as a building site?
4. A liquid limit of 50 sometimes is assumed to be a separation point between expansive and non-expansive clay. Should the criterion be different for floor loads and foundations?

5. You are asked to give an opinion regarding possible causes for cracks in walls and basement floors, and the house is on a hillside and has a walk-out basement. The hillside soil has an expansive clay subsoil or B horizon. Draw a cross section that can explain the problem.
6. A 1-m thick layer of sand is not as effective as the same thickness layer of non-expansive clay for preventing volume changes in an underlying expansive clay. Why?
7. Why is the designation "replacement method" inappropriate for control of expansive clay with a layer of non-expansive clay? Briefly explain how the control method may work.
8. Which are the following are Type G (geological) and Type P (pedological) expansive clays and what are the origins?
    a. Expansive shale, North Dakota.
    b. Expansive clay from glacial melting and deposited in Lake Agassiz, Saskatchewan, Canada.
    c. A buried "gumbotil" paleosol (ancient soil) layer of expansive clay separating two glacial deposits in northern Missouri.
    d. Loess soil deposited a hundred miles (150 km) from its source.
    e. Delta rice paddy, Viet Nam.
    f. Expansive claystone, Colorado.
9. The homeowner uses a garden hose to flush a gopher out of its boring. The house is on collapsible loess. Who is likely to be the winner?
10. What makes quick clay quick? Is quick clay likely to be expansive? Why or why not?
11. Is there any relationship between quicksand and quick clay?
12. How can deep dynamic compaction (DDC) prevent future liquefaction?
13. What is wrong with movie depictions of quicksand (a) in a desert and (b) when the villain sinks so only his hat remains floating?
14. Another approach to explain the replacement method for expansive clay uses capillary attractions (Houston et al., 2016). How may that be influenced by hysteresis in the retention of clay mineral interlayer water?
15. Heaved floors and footings and supported structures have been saved by creating a non-expansive clay sublayer with 4–5 inch (100 mm) borings every 4–5 feet (1.5 m) and filled with quicklime. How deep should the borings be?

References

Briaud, J-L., Abdelmalak, R., Zhang, X., and Magbo, C., "Stiffened slabs-on-grade on shrink-swell soil: New design method," *ASCE Journal of Geotechnical and Geoenvironmental Engineering*, 142(7):1–12, 2016. Summarized online: ibid., "Design of Stiffened Slabs-on-Grade on Shrink-Swell Soils.

Brinkgreve, R. B. J., "Automated Model and Parameter Selection," *Geostrata*, January/February 2019. ASCE GeoInstitute.

Chen, F. H., *Foundations on Expansive Clay Soils*, Elsevier, Amsterdam, 1985.

Handy, R. L., "Loess Distribution by Variable Winds," *Bulletin of the Geological Society of America*, 87:915–927, 1976.

Handy, R. L., "Geology, Soil Science, and the Other Expansive Clays," *Geotechnical News*, 20(1):40–45, 2002.

Houston, S. L., Bharadwaj, A., Welfert, B., Houston, W. H., and Walsh, K. D., "Unsaturated soil mechanics principles to remove and replace mitigation for expansive clays." *ASCE Journal of Geotechnical and Geoenvironmental* Engineering,. 142(4):04015102-1-11, 2016. doi 10.1061/(ASCE)GT.1943-5606.0001429.

Katti, R. K., Lad, H. B., Fotedar, S., and Kulkarni, S. R., "Depth Effect in Expansive Clays," *Proc. II Int. Res and Engg. Conf. on Exp. Clay Soils*, Texas A. and M. Univ., pp. 362–373, 1969.

Senich, D., Demirel, T., and Handy, R. L., "X-ray Diffraction and Adsorption Isotherm Studies of the Calcium Montmorillonite-H_2O System," *Highway Research Record*, 29:23–54, 1967.

Williams, W. W., and Handy, R. L., "Chemical Stabilization of an Active Landslide," *Civil Engineering*, August:62–65, 1967.

Prof. R. K. Katti (1928–2010): Discoverer of a widely used method for controlling expansive clay by covering with a layer of non-expansive clay.

CHAPTER 5

Stresses in Soils

Spreading Like Ripples on a Pond

5.1. Concentrated Stresses

A first step toward predicting settlement of a foundation involves estimating pressures created in the underlying soil, as stresses spread and decrease in intensity with distance from a point of application. In 1885, the French mathematician M. J. Boussinesq (1842–1929) presented a solution based on ideal elastic behavior for stresses created from a point load on the surface of solid mass. Boussinesq's theory is illustrated in Fig. 5.1. Contours of equal stress are circles. Since foundation loads are not point loads, additional steps are required to enable applications of the theory to foundations.

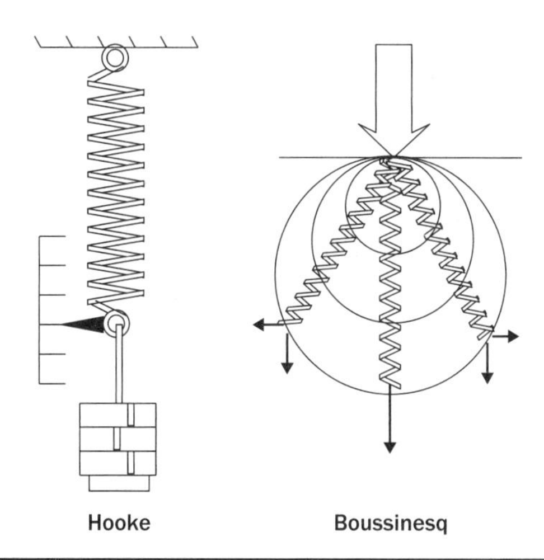

FIGURE 5.1. Hooke's Law, the basis for Boussinesq theory. (Image source: *Geotechnical Engineering: Soil and Foundations Principles and Practice* by Richard L. Handy and Merlin G. Spangler. © 2007, McGraw-Hill Companies.)

5.2. Adapting Boussinesq Theory

Vertical components of stress are most relevant for predicting settlement. The Boussinesq relationship was integrated for area loads by Prof. N. M. Newmark at the University of Illinois. Contours of vertical stress are shown in Fig. 5.2 and describe a *pressure bulb* that is different for linear and for square foundations, as the theoretical vertical stress under a linear foundation decreases to a 10 percent level at a depth that is about 6.4 times the foundation width, and the same reduction under a square foundation occurs at about 2 times the foundation width because the decrease is in all directions. This is an important consideration when both types of foundations are used to support the same structure, as a linear foundation extends stress about 3 times deeper and may encounter different soil layers.

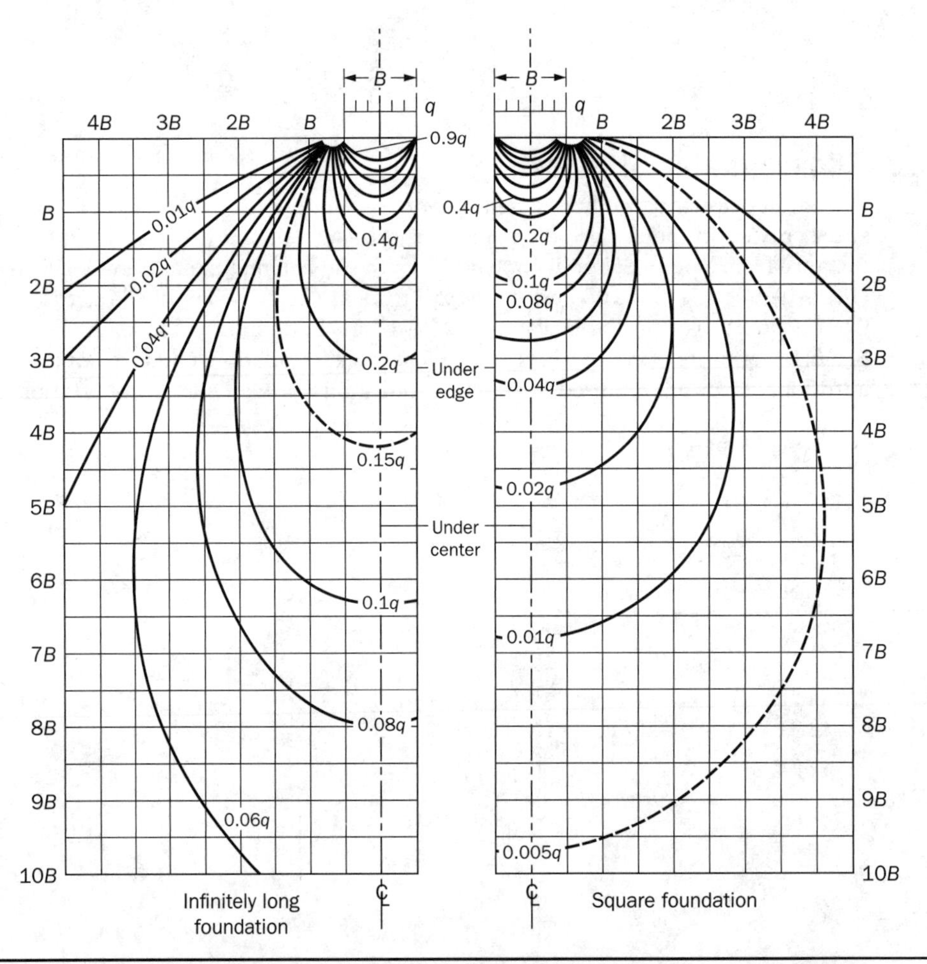

FIGURE 5.2. *Pressure bulbs*, contours of vertical stress in soil resulting from a surface area pressure q evenly distributed across an area of width B, based on ideal elastic behavior of the soil. (Adapted from Sowers and Sowers, 1961.) (Image source: *Geotechnical Engineering: Soil and Foundations Principles and Practice* by Richard L. Handy and Merlin G. Spangler. © 2007, McGraw-Hill Companies.)

Boussinesq's equation was integrated and extended to area loads by Prof. Nathan Newmark at the University of Illinois, and modified by H. M. Westergaard and Donald Burmister to apply to layered soils. It was integrated by M. G. Spangler to calculate horizontal pressures transmitted from surface loads to retaining walls and to buried pipes. Stress levels in soils underneath foundations not only can determine the amount of settlement but also are relevant to possible punching-shear bearing capacity failures discussed in the next chapter.

5.3. A Snag in the Relationship

The Boussinesq derivation famously assumes an "ideal, homogeneous, isotropic, elastic" material. As shown by test data in Fig. 3.7 and interpreted in Eq. (3.1), during normal consolidation the amount of compression is linear to the *logarithm* of the applied stress and not to the stress itself. Figure 5.3 shows that the modulus remains fairly constant during initial stages of loading and then the modulus increases linearly with pressure as a soil consolidates. The Boussinesq relationship, therefore, strictly applies only to the soil that is overconsolidated and behaves elastically. The stress contours shown in Fig. 5.2 do not incorporate the influence from increasing stiffness with depth, so they overestimate the influence depth for a foundation on normally consolidated soil. Correction is difficult because every soil behaves differently. One effect is the reduction in the penetration depth, and if an arbitrary cutoff is made at 10 percent of the foundation stress, the error cannot exceed 10 percent and probably will be much less than that.

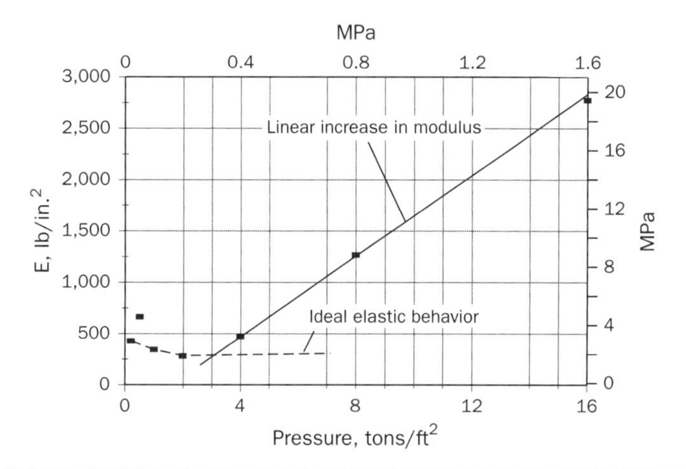

FIGURE 5.3. The increase in elastic modulus during normal consolidation is not included in a conventional settlement analysis. (Image source: *Geotechnical Engineering: Soil and Foundations Principles and Practice* by Richard L. Handy and Merlin G. Spangler. © 2007, McGraw-Hill Companies.)

5.4. Approximating the Pressure Distributions

Linear distributions of vertical stresses, as shown in Fig. 5.4, are convenient and are commonly employed, particularly for small structures. The estimates may be overly

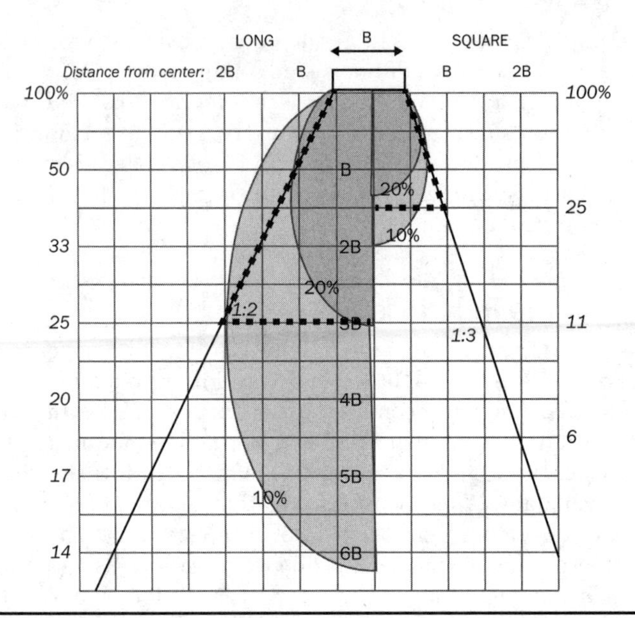

FIGURE 5.4. Linear approximations for influence coefficients at shallow depths, 1:2 for long and 1:3 for square or round surface area loads. Shaded areas show percentages from elastic theory with no increase in modulus, and heavy dashed lines show 75 percent reductions in foundation pressure at depths 3B and 1.5B using linear approximations.

conservative. Sides of squares in Fig. 5.4 are one-half of the foundation width, and Boussinesq distributions are shown by shaded areas. At a depth 3B under the center of a linear foundation the theoretical stress is 20 percent of the foundation stress and a linear distribution gives 25 percent. However, the linear distribution extends over a wider area. At a distance 2B from the center of the foundation it is 25 percent of the foundation pressure compared to 10 percent based on the elastic theory. The linear distributions, therefore, can exaggerate pressures created in soils under nearby existing foundations, and should not be used for this purpose.

A common procedure is to either identify soil layers and estimate foundation stresses in centers of the layers or divide the soil into arbitrary layers to estimate stresses. The amounts of consolidation of each layer then are calculated from the layer thicknesses and changes in void ratio determined from consolidation tests. If the layers are for different soils, a consolidation test will be required for each soil.

5.5. Preloading

An easy way to settle the settlement problem is to preload soil at a building site ahead of time to equal or exceed anticipated foundation loads and wait for the load to settle. As the waiting period obviously is critical, settlement should be monitored in time, or it can be based on experience if there is an assurance that the foundation soils are consistent. Usually the preload in a pile of soil then can be moved on to the next construction site. Some recommendations are as follows:

1. Anticipated foundation loads are likely to be concentrated under outer edges of a structure to support bearing walls; therefore, for maximum reliability the preload soil should be spread out and not just heaped up in the middle. (This often is ignored, but that does not mean it should be.)
2. Time must be sufficient to allow consolidation of the underlying soil. Since dry soil will resist consolidation, there may be a delay as water becomes available from consolidation of the underlying saturated and consolidating soil. It therefore is important to measure settlement and establish a time relationship.
3. Settlement can be monitored (a) through a conventional elevation survey of a stable benchmark to temporary benchmarks placed on top of the preload, or (b) by installing steel pipes or rods to bear on bedrock or flat steel plates installed prior to filling; therefore, settlement is measured from the lengths extending above the surface of the completed fill. Additional lengths are added as necessary to stay above the fill, and should be flagged to avoid being run over by earth-moving equipment.
4. As settlement should slow down when the system approaches equilibrium, the rate may be proportional to the amount of departure from a final equilibrium. A prediction then can be made using a *first-order rate equation,* described in Chap. 3.
5. Settlement should be measured frequently with preset time intervals to assure data continuity, with no sudden changes or irregularities.
6. *Role of soil shear strength and pore water pressure: Preloading should only be used where there is an assurance that the soil effective stress shear strength will not be exceeded and allow a bearing capacity failure.* This criterion applies to any foundation investigation and is discussed in the next chapter. (Preloading also can serve as a test for the soil bearing capacity.)

Example: The following data are obtained from settlement measurements of a preload placed on top of an old landfill. Use FORE (a first-order rate equation) to estimate the time required for the remaining amount of settlement to be less than 0.5 in. (12.5 mm).

Data: Settlement after 1 month is 5.5 in. (132 mm); after 2 months, 7.75 in. (160 mm); after 3 months, 8.5 in. (173 mm).

Solution: The FORE equation is: $\log(S_u - S) = at + b$, where $(S_u - S)$ is the departure from a final settlement, S_u, t is the time, and b is a constant. A graph is prepared so that trial values of S_u can be tested for the equation to provide a highest value of the statistical R^2, which should approach 1.0:

Data		Trial Calculations		
t (months)	**S (in.)**	**S_u (in.)**	**$S_u - S$**	**$\log(S_u - S)$**
12	5.5	9.5	4	0.60206
18	7.75	9.5	1.75	0.243038
22	8.5	9.5	1	0

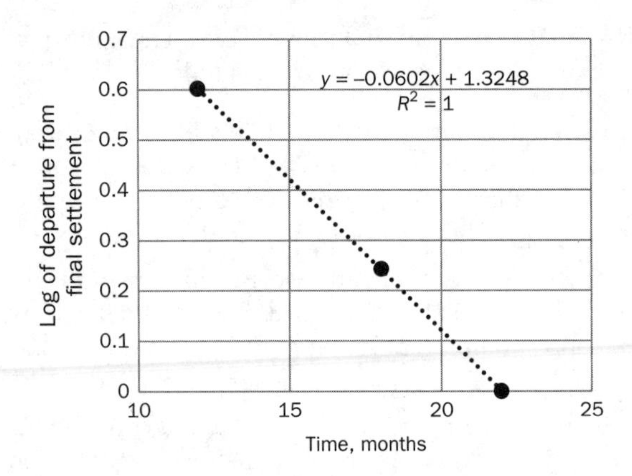

The trial end value of 9.5 in. gives the highest R^2 and straightest line. The time for 0.5 in. remaining settlement with the projected $S_u = 9.5$ in. is

$\log(9.5 - 9) = -0.0602t + 1.3248 = -0.301$

$t = 27$ months.

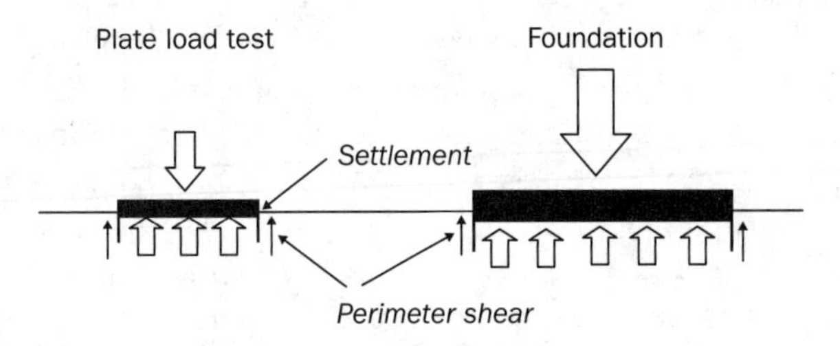

FIGURE 5.5. The role of perimeter shear is exaggerated in a plate bearing test compared to that in a large foundation.

5.6. A Plate Bearing Test as a Model Foundation

The plate used in a plate bearing test causes stress to be more concentrated by *perimeter shear* around the edges. This effect is illustrated in Fig. 5.5. The smaller the loaded area, the larger the role played by perimeter shear. This is significant when applying results from a plate bearing test to a much larger foundation contact area.

Prof. W. S. Housel (1901–1978) at the University of Michigan suggested a simple equation to isolate the contributions from contact pressure and perimeter shear:

$$W = P_m + A_n \tag{5.1}$$

where W is the applied load, P is the plate perimeter, A is the contact area, and m and n are constants. Dividing by A gives

$$\frac{W}{A} = \frac{P}{A}m + n \tag{5.2}$$

As the equation has two unknowns, *m* and *n*, it can be evaluated by conducting two plate bearing tests with two different sizes of plates. Note that *m* and *n* are not dimensionless, so units must be consistent. The resulting equation then can be applied using the same values of *m* and *n* for a full-size foundation.

Example: Plate load tests were conducted at an anticipated foundation depth with two different diameters of plates. One had a diameter of 2 ft (0.61 m) and the other of 3 ft (0.91 m). The following test results were obtained for 1 in. (25 mm) settlement:

	Plate 1	Plate 2
$A =$	3.14 ft^2	7.06 ft^2
$P =$	6.28 ft	9.42 ft

Solution:

	Plate 1	Plate 2
$W =$	2,820	5,600 lb
$W/A =$	898	793 lb/ft^2
$P/A =$	2	1.33 /ft

$$W/A = (P/A)\,(m + n)$$

Plate 1	Plate 2
$898 = 2m + n$	$793 = 1.33m + n$

$$n = 898 - 2m = 793 - 1.33m$$

$$0.67m = 105;\ m = 157$$

$$n = 898 - 2(67) = 734$$

For a square footing 5 ft on a side, $A = 25\ ft^2$; $P = 20$ ft:

$$W/25 = (20/25)\,(149) + 510 = 704$$

$$W = 17{,}600 \text{ lb (8.8 tons)}$$

The calculated values of *m* and *n* can be substituted in Equation (5.2) to obtain perimeter shear for a particular size foundation on the same soil.

The contribution from perimeter shear decreases with increasing foundation size. However, the depth of the pressure bulb simultaneously *increases*, and if it extends down into stiffer soil, as is often the case in normally consolidated soil of river floodplains, it may compensate for the effect of perimeter shear. As an evaluation of perimeter shear requires two bearing tests on separated soil areas it often is ignored.

5.7. Performing a Plate Bearing Test

The setup for a field plate bearing test is illustrated in Fig. 5.6 and the steps are listed below.

1. An excavation is made to the anticipated foundation depth and smoothed with a layer of sand, and the steel plate placed and tamped down on the sand. Sides of the excavation may be cut back, if necessary. Plates may be stacked to minimize warping.

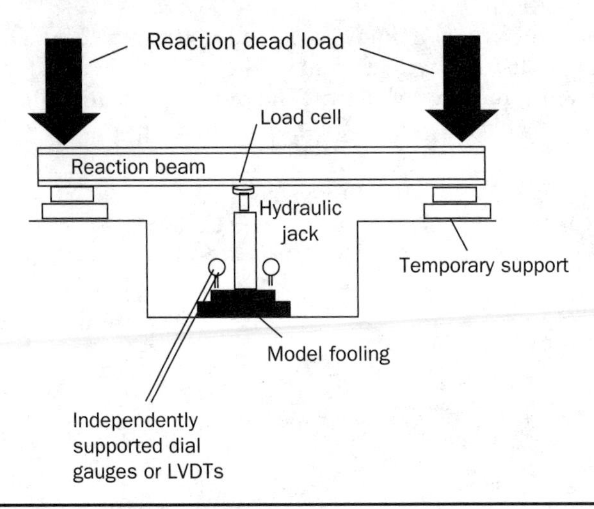

FIGURE 5.6. Schematic of a plate bearing test.

2. Loads are applied by jacking against a crossbeam that can be held down at the ends by the weight of heavy construction equipment, or by piles or helical screw anchors.
3. The applied load is monitored with a load cell placed between the jack and the beam, or from oil pressure in a calibrated hydraulic jack. Settlement is measured with a dial or electronic device that is independently supported on a crossbeam. Loads are applied incrementally to equal or exceed maximum anticipated foundation pressures. The installation should be shaded to prevent warping from heat of the sun, and protected from rain and from runoff water.
4. A critically important precaution is illustrated in Fig. 5.9 and discussed in the next section: *the larger the loaded area, the deeper the stress penetration.* A plate load test may be unsuitable for layered soils unless the test site is excavated so that the test can be repeated at different depths.

Plate bearing tests are expensive and time-consuming but can be used to confirm a design or aid in resolving conflicting results from other test methods. A similar loading arrangement is scaled up for pile load tests.

5.8. The Progressive Nature of a Bearing Capacity Failure

The downward progression of perimeter shear during loading was analyzed by O. K. Fröhlich in 1934 (Jumikis, 1969), based on the elastic theory. It is illustrated in Fig. 5.7. As one side inevitably will be weaker in shear than the other, a bearing capacity failure usually involves tilting (Fig. 5.8). The analysis of bearing capacity is presented in Chap. 6.

5.9. Plate Bearing Tests on Weathered Soil Profiles

Figure 5.9 illustrates how a plate load test can distort results to the unsafe side in layered soils. It is not uncommon for a dense, stiff clayey subsoil ("B" horizon) layer to

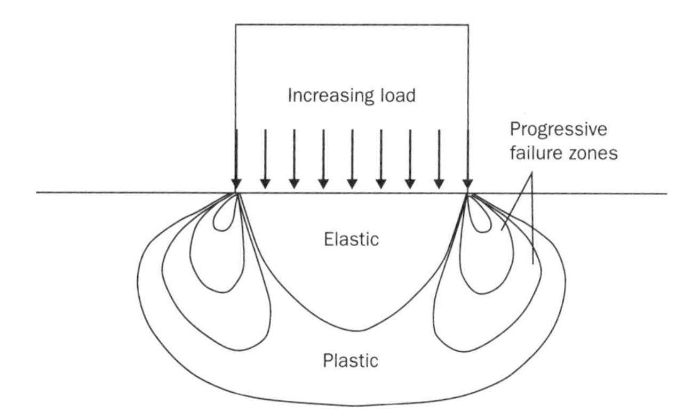

Figure 5.7. Downward extension of perimeter shearing stresses can lead to a bearing capacity failure. (From the analysis of Fröhlich based on the elastic theory (Jumikis, 1969).)(Image source: *Geotechnical Engineering: Soil and Foundations Principles and Practice* by Richard L. Handy and Merlin G. Spangler. © 2007, McGraw-Hill Companies.)

Figure 5.8. A classic bearing capacity failure that occurred prior to recognition of the role played by pore water pressures. The bins were supported, jacked up, and put into service. (Image source: Transcona Historical Museum, Canada.)

occur on top of a weaker soil. Before a plate load test is conducted, the soil is excavated so that the plate can bear on the weaker soil. Plate bearing strength also is affected by moisture content; the drier the soil, the larger the resistance, and freezing will have an obvious effect on test results.

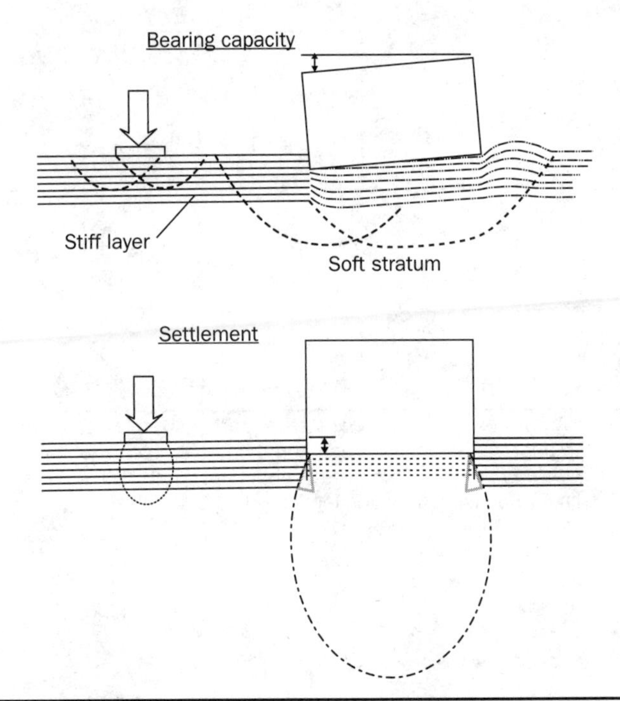

FIGURE 5.9. Even with a perimeter correction, plate bearing test data on layered soils can give results that are on the unsafe side for both bearing capacity (top figure) and settlement.

5.10. Foundation Stresses Transferred to Nearby Unyielding Walls

The above discussion relates only to vertical stress, but as shown by the arrows in Fig. 5.1, vertical pressure also creates lateral stress that can be transmitted to nearby walls. Stress measurements by Prof. M. G. Spangler (1936) at Iowa State College gave an unexpected result, that lateral pressures from a surface load are almost exactly *doubled* when imposed against an unyielding wall—because the wall is unyielding.

Figure 5.10 shows lateral pressures against a wall calculated from elastic theory, and the closer the foundation is to a wall, the larger the amount of lateral pressure transmitted to the wall. A separation distance equal to the width *B* of a foundation still can result in 40 percent of the foundation pressure being transmitted as horizontal stress to the wall, and if the wall goes, the foundation is likely to follow.

An option is to extend a foundation downward to the same elevation as the wall foundation, but excavation for the foundation will remove overburden pressure that, as discussed in the next chapter, may be an important factor contributing to the safety of the wall.

A wall that is safe from a bearing capacity failure still can be affected by the bulb of pressure created by a nearby foundation. As shown in Fig. 5.2, the bulb of vertical stress can extend outward and affect pressure on soil underneath the wall and contribute to possible settlement, a matter of considerable importance if the wall is supporting another structure.

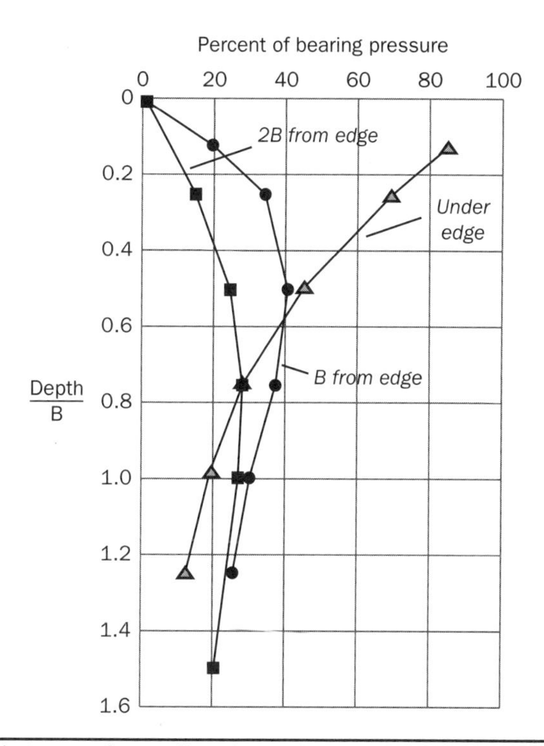

FIGURE 5.10. Lateral stresses from soil against a rigid wall at various depths and distances from the foundation. Stresses calculated from the elastic theory and doubled by a mirror effect from a rigid wall.

A *common option* is to support a new structure on a deep foundation such as piles in order to transfer stress downward and mitigate a possible adverse effect on an adjacent wall.

Regardless of the option that is selected, *a crack survey with photographs of the existing structure should be mandatory, and usually can be conducted with permission of the owner.*

5.11. Strength Gains from Aging

Soil is an open system, and numerous investigations have revealed significant gains in soil strength upon aging (Mitchell and Soga, 2005). The gain usually can be attributed to secondary cementation at grain contacts. The cement can be with concentrations of clay mineral particles that are relatively mobile, or by secondary deposits of calcium carbonate, silica, and other salts. A slow improvement in contact bonding is supplemental to the adhesion theory for friction. Schmertmann (1991) notes strength gains as high as 50–100 percent can result from aging, even in sand.

Effects from aging can be evident in laboratory strength tests of compacted soil samples as they recover strength with time. As strength increases, the seismic velocity significantly may decrease, suggesting a realignment of water molecules to optimize electrical attractions and cushioning contacts between clay particles. Even though the strength increases, the seismic velocity measured in compacted clay tends to decrease and stabilize after a day or two, probably as clay particles become more uniformly separated by films of water.

Interruptions during Pile Driving

Time-related strength gains can be a nuisance during pile driving because if for any reason driving is interrupted, the pile may "freeze" and will not move when driving is resumed. The most likely cause is time-dependent dissipation of excess pore water pressure that has aided the soil resistance to penetration by a pile.

5.12. A Convenient Maximum Depth for Pressure Calculations

Pressure from a surface load theoretically is asymptotic with increasing distance, but there are practical limits for the calculations. A common assumption is that the influence from a foundation pressure becomes negligible below a depth defined by $0.25q$, where q is the foundation pressure. A more conservative estimate is at $0.10q$. Both are shown by shaded areas in Fig. 5.4. A common acceptance level for settlement is 1 in., which is not only convenient but also easy to remember.

Problems

1. Usual applications of the elastic theory to predict foundation settlement assume a constant modulus. How may that be influenced by consolidation of the soil? How will consolidation affect the depth of the pressure bulb? Will it tend to increase or decrease the predicted amount of settlement?
2. What is secondary consolidation, and how may it affect long-term settlement calculations?
3. Assume that wall and column footings for a structure are designed to exert the same soil contact pressure in order to assure uniform settlement. Is that assumption correct? Why or why not?
4. An alluvial floodplain clay has no history of prior loading, but consolidation tests nevertheless reveal a substantial preconsolidation pressure. Why? Is a plate bearing test reliable on this clay? Why or why not?
5. What is the significance of mixed gray and brown colors in soil above a groundwater table?
6. When a building excavation is dewatered with perimeter wells, cracks are observed in walls in an adjacent building that is supported on shallow foundations. You are approached by the owner of the damaged building. What should be your approach to the problem? Might the use of crack-monitoring gauges provide information that could be relevant in a lawsuit? Can anything be done to remedy the situation?
7. A clay soil has a preconsolidation pressure of 1 Tsf (0.1 MPa) and $C_c = 0.21$. Use Fig. 5.2 to estimate the load on a 1 m^2 foundation that will cause 25 mm (1 in.) of settlement.

8. In the preceding problem, estimate an appropriate footing size to carry a 2.5 ton/ft (350 N/m) wall load.
9. In the example on P. 85, estimate the load for 1 in. of settlement of a 4 ft square foundation based on perimeter shear.

References

Jumikis, A. R., *Theoretical Soil Mechanics*, Van Nostrand Reinhold, New York, 1963.

Mitchell, J. K., and Soga, K., *Fundamentals of Soil Behavior,* 3rd ed., John Wiley & Sons, Hoboken, NJ, 2005.

Schmertmann, J. B., "The Mechanical Aging of Soils," *Journal of Geotechnical Engineering*, 117(9):1288–1330, 1991.

Sowers, G. B., and Sowers, G. F., *Soil Mechanics and Foundations,* 2nd ed., The Macmillan Co., New York, 1961.

Spangler, M. G., "The Distribution of Normal Pressure on a Retaining Wall Due to a Concentrated Surface Load," *Proceedings of the International Conference on Soil Mechanics and Foundation Engineering*, Graduate School of Engineering, Harvard University, Cambridge, MA, pp. 200–207, 1936.

J. V. Boussinesq (1842–1929), the French physicist best known in geotechnical engineering for his solution for vertical stress from a point load at the surface of an elastic medium.

CHAPTER 6

Evaluating Soil Shear Strength

There's Many a Slip Unless Something Is Done about It

6.1. Bearing Capacity and Settlement

The transition from foundation settlement to a foundation bearing capacity failure is like comparing an automobile crash with a leaky tire. Although some settlement usually is acceptable and almost inevitable (Fig. 6.1B) and should, bearing capacity failures are rare and occur when stresses imposed on soil exceed the soil shear strength (Fig. 6.1A).

6.2. Friction

In 1493 while Columbus was discovering America, Leonardo da Vinci was discovering friction. He made a sketch that shows a string pulling a stack of wood blocks across a surface, and his sketch indicated that the higher the stack, the larger the amount of pull that is required to cause sliding. That apparently satisfied his curiosity, so he went back to paying jobs. The sketch was discovered in 2016 by Prof. Ian Hutchings of the University of Cambridge.

Two centuries later the relationship between friction and normal stress was rediscovered by a French instrument maker Guillaume Amontons (1663–1705) and is called Amontons' First Law:

$$F = N \times a \text{ coefficient of friction} \tag{6.1}$$

This can be counterintuitive: Is the force the same to slide a brick standing on end as it is for the same brick laid flat on the same surface?

Answer: Yes.

A coefficient of sliding friction normally varies between about 0.2 and 0.5 unless the surfaces are separated with a lubricant or involve sandpaper.

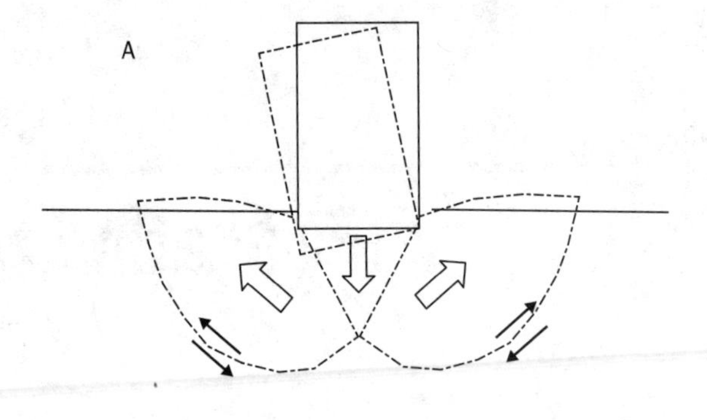

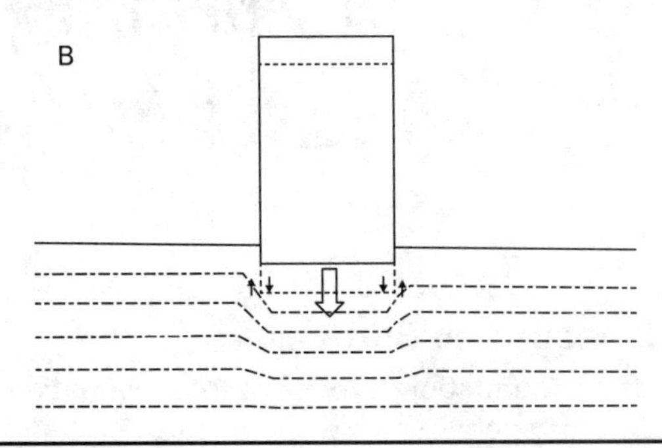

FIGURE 6.1 (A) A bearing capacity failure involves sudden shearing of underlying soil, and may cause tilting because of uneven weight distribution or varying soil strength properties. (B) Settlement involves compression of the supporting soil. It is slow and inevitable, and should slow down as the underlying soil compresses.

A graph with N values along the horizontal axis and F values stepped off vertically gives a straight line with a slope that is the *friction angle*, ϕ. Hence

$$F = N \tan \phi \tag{6.2}$$

This is not a difficult equation, as only one measurement of F and N is required to establish a coefficient of friction and $\tan \phi$. When applied to soils ϕ is the *angle of internal friction.*

Friction Angle and Slope Angle

The friction angle ϕ is readily demonstrated with a brick on a tilted board, or the slope angle for a pile of dry sand, where it is the *angle of repose.* This is illustrated in Fig. 6.2.

Amontons' Second Law

Dividing both sides of Eq. (6.2) by the contact area, A, gives:

$$\frac{F}{A} = \left[\frac{N}{A}\right] \tan \phi \tag{6.3}$$

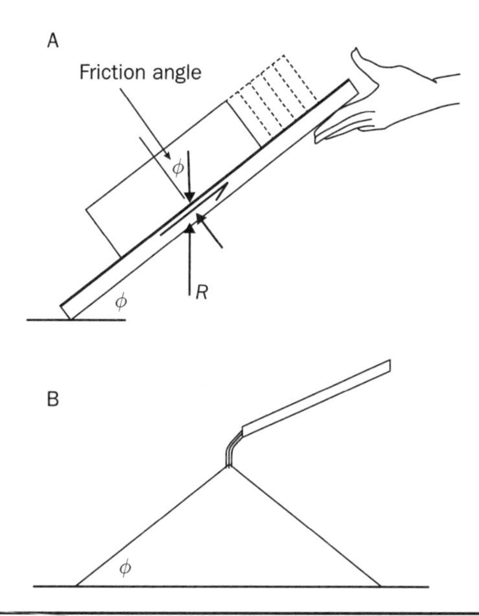

Figure 6.2 (A) The two ϕ angles in the upper figure can be shown to be equal through mutual perpendiculars. (B) ϕ also is the *angle of repose* for loose sand. (Image source: *Geotechnical Engineering: Soil and Foundations Principles and Practice* by Richard L. Handy and Merlin G. Spangler. © 2007, McGraw-Hill Companies.)

where F/A is a *shearing stress* that is required for sliding, and is proportional to the *normal stress, N/A*. The Second Law allows the introduction of pore water pressure:

$$\frac{F}{A} = \left[\frac{N}{A} - u\right]\tan\phi \tag{6.4}$$

where $\left[\frac{N}{A} - u\right]$ is the *effective* normal stress.

The Greek Connection

Equation (6.3) can be rewritten with Greek letters τ_f (tau sub f) for F/A and σ (sigma) for N/A:

$$\tau_f = (\sigma - u)\tan\phi \tag{6.5}$$

or

$$\tau_f = (\sigma')\tan\phi \tag{6.5a}$$

where σ' is the normal *effective* stress. This is the expression commonly used in geotechnical/foundation engineering.

Coulomb's Equation

Charles Coulomb (1736–1806) was a French military engineer who observed that clay can stand in a vertical slope, so something must be left out of Eq. (6.5).

He suggested that there must be a cohesion component that does not depend on normal stress. This observation is expressed by

$$\tau_f = \sigma \tan \phi + c \tag{6.6}$$

where c is a cohesive component of shear strength, commonly called *cohesion*. Based on his experience in the military, Coulomb used static equilibrium to develop a formula for soil stresses on retaining walls. It is an oversimplification because it assumes block behavior, but is easy to program and remains widely used today.

Coulomb retired from the military service for health reasons, and entered a contest supported by the King to develop a more accurate marine compass for ocean navigation. Although he did not win the contest, he invented the *torsion balance* that led to bigger things and still is a common constituent of electronic weighting devices. Static electricity became a novelty after Benjamin Franklin nearly electrocuted himself, and Coulomb used his sensitive balance to measure attraction and repulsion forces between electrically charged pith balls. He anticipated a linear relationship to separation distance, but it turned out to be the *square* of the distance and now is known as *Coulomb's Law.* It establishes the orbits of satellites.

6.3. Friction Angle in Soils

Friction between grain contact surfaces in soils involves more than just sliding, as closely packed soil grains must slide up and over one another. That adds a volumetric component to friction, *dilatancy*, and the denser the soil, the larger the contribution from dilatancy. This was suggested in 1885 by the Irish scientist Osborne Reynolds (1842–1912), a professor at the University of Manchester. Reynolds' major work was in fluid mechanics, where he popularized *Reynolds number* that is used to differentiate between viscous and turbulent flow.

The dilatant contribution to sliding friction ends as soon as soil particles become rearranged to allow the development of a continuous slip surface. As the dilatant component of friction is lost, sliding can continue – until a landslide runs out onto a more level surface, or a foundation sinks into the ground far enough that the weight of the displaced soil prevents further sliding.

Dilatancy in Design

Design normally is based on the maximum developed friction that includes dilatancy; a *factor of safety* then is applied to make sure that the loss of a dilatant contribution to shear strength never happens.

6.4. A Direct Shear Test

A simple method for measuring soil shear strength is illustrated in Fig. 6.3. Normal stress is applied vertically and held constant with a piston and regulated gas pressure or a weight and lever system, and shearing stress is increased horizontally until the soil shears along a horizontal surface. The test sample usually is similar in size and shape to that used in a consolidation test. Drainage of pore water pressure is allowed, and pore water pressure is not measured. The test is conducted slowly to allow drainage. Shearing force is slowly increased and monitored until the soil shears. In a typical test,

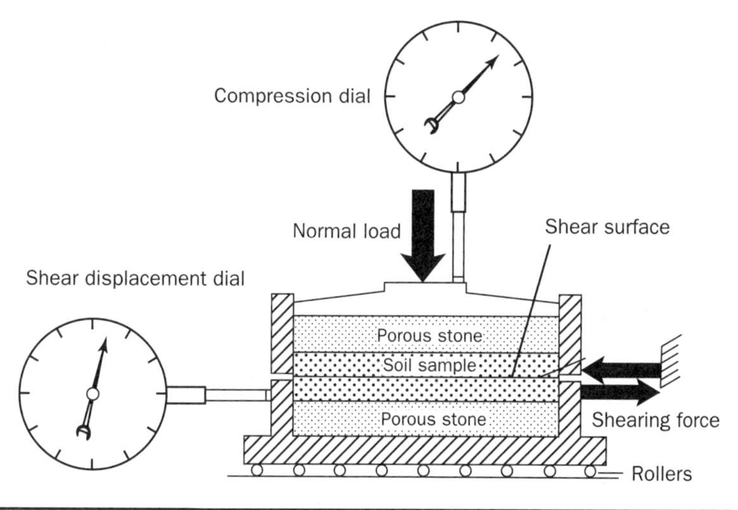

FIGURE 6.3 Schematic of a direct shear test apparatus.

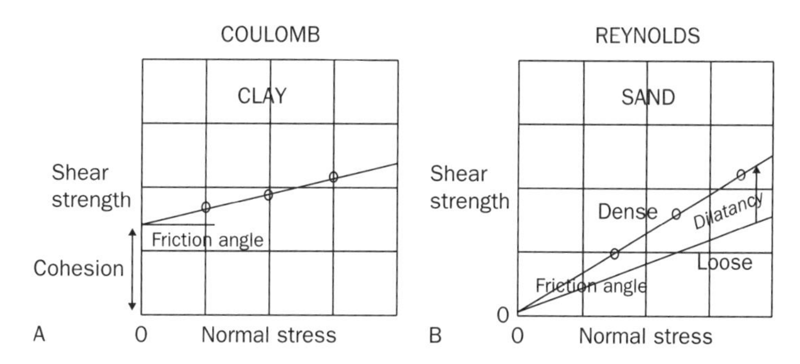

FIGURE 6.4 Direct shear tests can confirm (A) Coulomb's theory for two components for soil shear strength and (B) Reynolds' theory for the role of dilatancy.

shearing resistance increases to a peak and then declines as the dilatant component of friction is lost.

A direct shear test usually is conducted three times on three samples of the same soil, and data graphed (Fig. 6.4). The slope of the line is the *angle of internal friction*, and the intercept on the y-axis is the *cohesive shear strength*, usually referred to by geotechnical engineers as simply "cohesion." As no pore water pressure is allowed in this test, *effective stress shear strength parameters* are directly measured and can be designated as c' and ϕ'.

Influence of Layering

Soils deposited from wind or water usually show some horizontal layering or stratification. Since shearing is in the direction of layering the use of different test samples affects variability and hence reliability of the test results, direct shear tests have largely been replaced with other tests. In a *simple shear test* the test specimen is confined between parallel steel rings.

The Borehole Shear Test (BST)

The Borehole Shear Test shears soil vertically in the perimeter of a borehole, so shearing cuts across and tends to average the influence from layering. As in a laboratory direct shear test, when properly performed drainage permits direct measurements of effective stress soil internal friction and cohesion. Drainage times are cumulative, with 15 minutes usually allowed for the first data point and 5 minutes for each successive data point. A diagram of the apparatus is depicted in Fig. 6.5.

The BST also is a *stage test,* which means that it is conducted with successively higher values of normal stress on essentially the same soil. That is permissible if after each episode of shearing, a higher applied normal stress reconsolidates the sheared soil so it becomes stronger and the shear surface moves outward to engage undisturbed soil. Evidence that this occurs includes (a) creation of a compact layer of soil that builds up on the grooved shear plates, and (b) successive data points that inevitably reveal an increase in the shearing resistance. Strength would decrease if shearing were to be repeated in the same soil.

The test is conducted in 75 mm (3 in.) borings made by pushing a thin-walled (Shelby) steel tube, or shallow tests can be conducted in hand-augered holes. Soils that

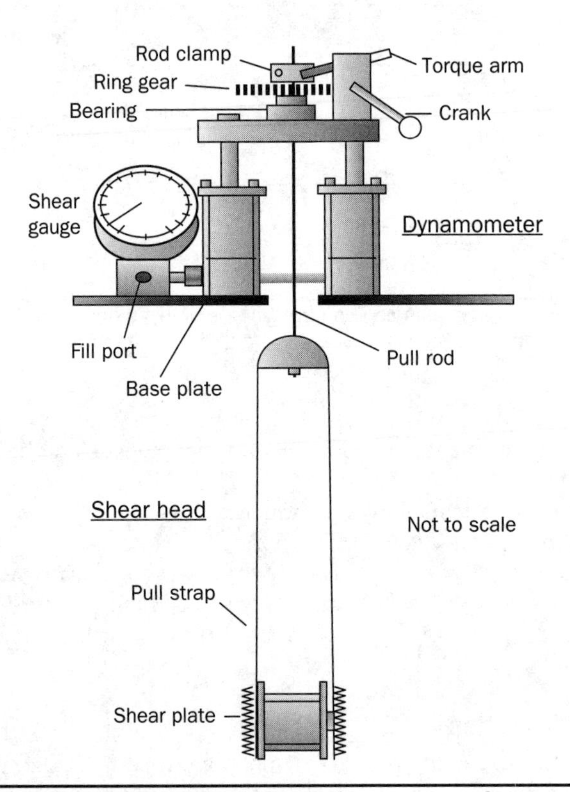

Figure 6.5 Schematic of the Borehole Shear Tester. (Image source: Courtesy of Handy Geotechnical Instruments, Inc.)

do not hold an open boring are tested through hollow-stemmed auger, and shear plates push aside a mud cake. Pore pressure can be monitored, but a simpler procedure is to increase consolidation times between data points and determine if that affects the data. Example graphs are shown in Figs. 6.6 and 6.7. As tests are conducted on-site, they can

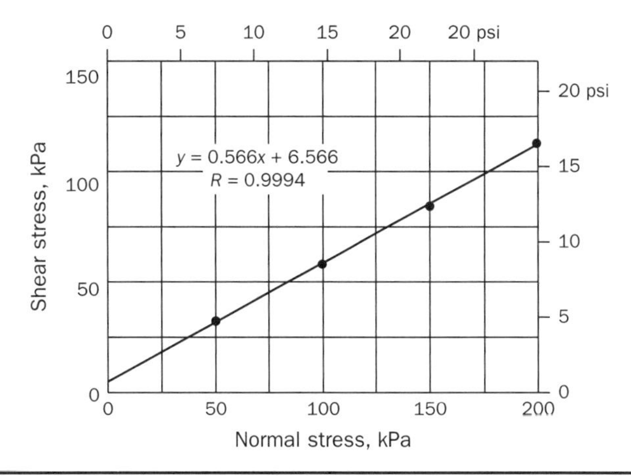

FIGURE 6.6 BST data from decomposed granite that did not yield intact samples for laboratory tests. (Data courtesy of National Geotechnical Consultants, The Gap, Queensland, Australia.)

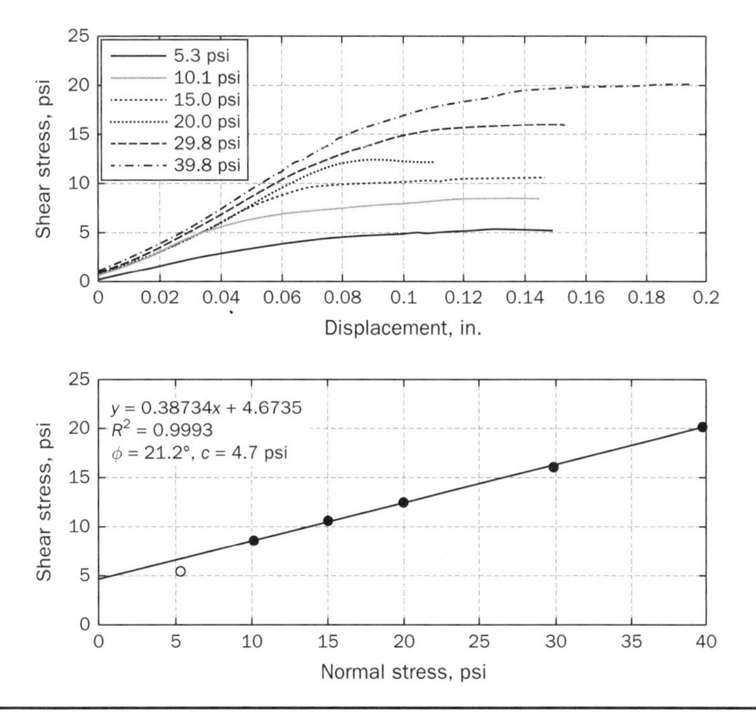

FIGURE 6.7 The fully automated Borehole Shear Test eliminates the human factor, and high R^2 values confirm Coulomb's speculation with regard to two components to soil shear strength, internal friction and cohesion. (Data courtesy of Dr. Jeremy Ashlock, Iowa State University.)

indicate if additional testing is needed. A high-pressure version is used to measure the strength of sedimentary rocks and avoid bias to the unsafe side by only testing intact samples, and is used to test foundation rocks under large dams.

6.5. Unconfined Compression Test

The unconfined compression test is used to test Portland cement concrete and has limited use in soils. It is not suitable for loose sand, and usually is assumed to be undrained when testing saturated clay, with all applied stress going to pore water pressure. That is not supported by visual evidence because water normally does not squeeze out of a test specimen. However, as depicted in Fig. 6.8, the assumption of zero internal friction is on the safe side for design. Test specimen lengths are at least two times the diameters in order to avoid influence from end friction.

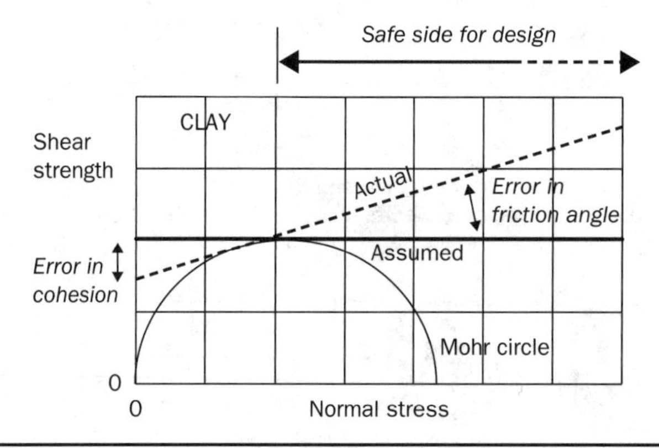

FIGURE 6.8 The unconfined compression strength (plotted at the right edge of the Mohr circle) with an assumed friction angle of zero generally is regarded as being on the safe side for design.

6.6. Mohr's Theory

In 1882 Prof. Christian Otto Mohr (1835–1918) at Stuttgart Polytechnic in Germany devised a graphical method to determine normal and shearing stresses on any inclined surface if major and minor principal stresses are known. The solution is embodied in *Mohr's circle,* which is fundamental in geotechnical engineering. It can apply to the unconfined compression test, as shown in Fig. 6.8, and also to a *triaxial shear test* in which lateral pressures are increased to simulate existing and anticipated field conditions. In a triaxial test a cylindrical soil specimen is enclosed in a rubber membrane and enclosed in a pressure chamber so that it can be subjected to a controlled lateral stress. The test apparatus is shown in Fig. 6.9.

In a conventional triaxial test the lateral stress is held constant as the specimen is axially loaded to failure. Two soil specimens that were tested to failure are shown in Fig. 6.10, and reveal shear failure surfaces that are inclined at an approximate angle of $45 + \phi/2$ with horizontal, which is in accordance with Mohr's theory. This construction is shown in Fig. 6.11.

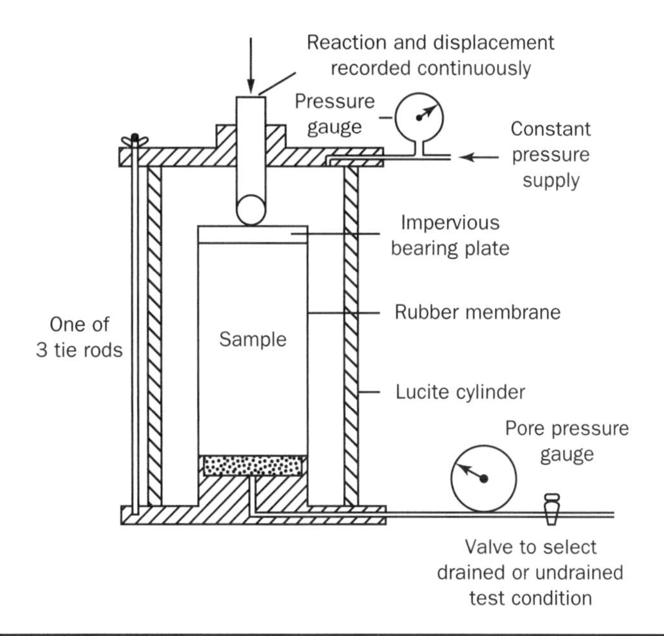

FIGURE 6.9 Schematic diagram of a triaxial shear test. If water is used to apply lateral stress and fill the gap between the soil sample and the cylinder, volume changes in the soil can be monitored from displacement of the water. (Image source: *Geotechnical Engineering: Soil and Foundations Principles and Practice* by Richard L. Handy and Merlin G. Spangler. © 2007, McGraw-Hill Companies.)

FIGURE 6.10 Silty clay soil specimens after triaxial compression testing. Shear failure angles can be seen in both specimens. (Image source: *Geotechnical Engineering: Soil and Foundations Principles and Practice* by Richard L. Handy and Merlin G. Spangler. © 2007, McGraw-Hill Companies.)

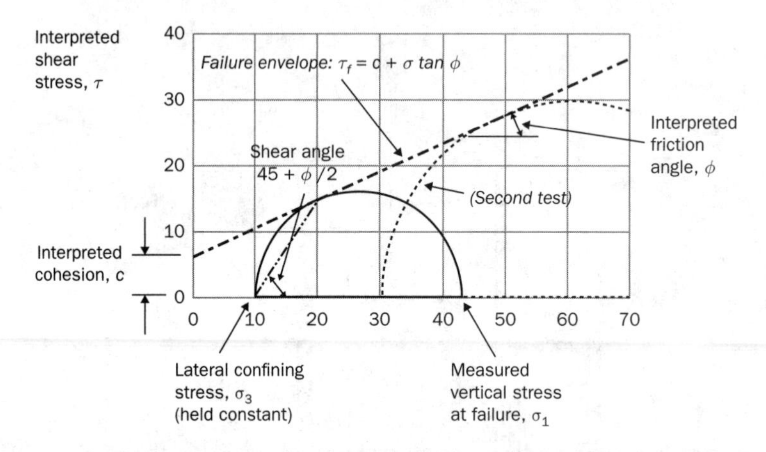

FIGURE 6.11 Two Mohr circles obtained from tests conducted with two different lateral confining stresses can be adequate to define a failure envelope with a cohesion and internal friction. Because of the sample variability, three tests usually are performed and can define an average failure envelope.

Although in a direct shear test the failure envelope goes through shearing failure stresses that define data points, in the interpretation of triaxial shear tests the equivalent data points are determined by a common tangent to Mohr circles. This construction is shown in Fig. 6.11, where the inclination of the failure surface is shown by the dash-dotted line on the left circle.

Pore Water Pressure

According to Terzaghi's theory of effective stress, pore water pressure pushes soil grains apart and subtracts from applied normal stress. An advantage of direct-shear types of tests is that drainage is allowed and specimens thin, minimizing the influence from excess pore water pressure. In addition, in Borehole Shear Tests drainage times are cumulative.

In triaxial shear tests the specimen is sealed but pore water pressure can be continuously monitored, usually through a filter located in the bottom platen. Subtracting the pore water pressure, u, from the lateral and vertical stresses converts them to *effective principal stresses*, designated by σ_3' and σ_1'. The common tangent then gives c' and ϕ' on a drained or effective stress basis.

6.7. A Difficult Problem

Triaxial shear tests rank high academically, but when applied to foundation design can be seriously compromised by sample variability, as a different soil specimen is used for each test. Some hypothetical and real examples are shown in Fig. 6.12. These examples can create an impossible situation for drawing a common tangent to Mohr circles, and additional tests of different specimens may only confirm the variability. With horizontal direct shear tests the difficulty can be amplified by horizontal layering, whereas vertical shearing promotes averaging.

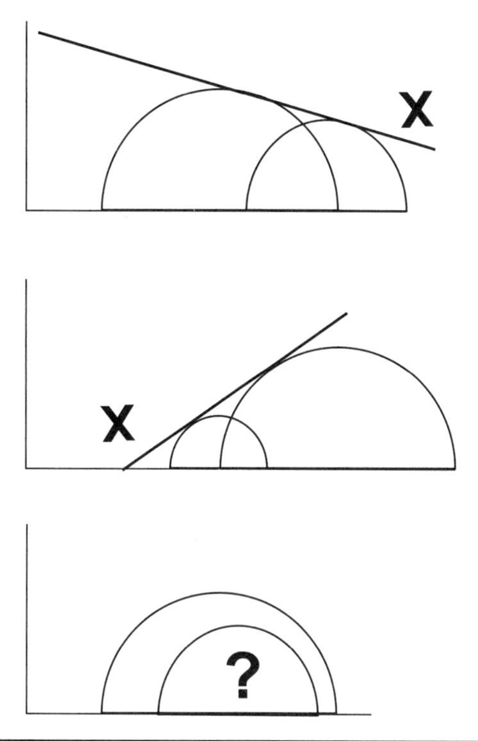

FIGURE 6.12 A practical problem of interpretation of triaxial shear test data: Test specimen variability and conflicted Mohr circles.

Stage Testing

Specimen variability can be countered by repeatedly testing the same soil specimen. This procedure can be employed in triaxial shear testing by *stage testing*, taking a specimen close to failure with a constant applied lateral stress, then as the stress–strain curve levels out, increasing the lateral stress to the next level and repeating the procedure. As some particle reorganization is likely to occur, the friction angle may be underestimated, which is on the safe side for design.

Lambe's Stress Path Method

Prof. T. William Lambe at MIT modified the triaxial test procedure to more closely simulate field loading conditions (Lambe and Whitman, 1979). Instead of taking a test to failure, a single test specimen can be used from each depth to predict foundation settlement. The lateral confining stress is estimated from the Jáky equation, $K_o = 1 - \sin \phi'$. That equation approximately applies to normally consolidated soil where settlement is most likely to pose a problem.

What about the Intermediate Principal Stress?

In a triaxial shear test the intermediate principal stress equals the minor principal stress, which according to Mohr's theory should make no difference. However, it allows a failure surface to be oriented in the most favorable direction. "Plane strain" tests with a rectangular soil sample held between two unyielding plates indicate a difference of the order of 1–2 degrees in the measured friction angle; therefore, this is not a major factor.

6.8. Statistical Analysis of Test Data

Conventional practice is to graph a controlled *independent variable* on an x-axis versus a measured *dependent variable* on the y-axis. A *linear regression* of y on x is readily performed with a computer to establish a linear equation and trendline.

Statistical *confidence limits* also can be calculated that enclose 90 percent or more of the existing and anticipated data. Confidence limits are used in process engineering. For example, if a batch of steel falls outside of the accepted limits, it can readily be incorporated into the next batch. That is not the case in foundation engineering. It gets one try. A graph of confidence limits also has a bowtie shape that exaggerates the variability of the cohesion intercept. Even confidence limits that enclose 99 percent of the existing and expected data leave a 1 percent probability that values will fall outside of those limits, and present a likelihood of failure of 0.5 percent that is not acceptable. The usual alternative is to in effect overdesign by incorporating a *factor of safety* of 3 or more, in order to ensure that a design will be on the safe side. There also can be a level of redundancy, as overstressing in one area can be compensated by a transfer of stress to an adjacent area that retains an acceptable factor of safety.

R^2 (*R* squared)

An R^2 value is readily calculated and printed on x-y graphs, and gives a measure of precision. The maximum value for R^2 is 1.0, if all data points are exactly on line. (The minimum value is –1.0 for an inverse relationship.)

In geotechnical and foundation engineering the main variability is in the material being tested, but variability also can be introduced by the test method. R^2 values in a range 0.98–1.0 can be expected when essentially the same material is repeatedly being tested, as shown in Fig. 6.7 for a representative Borehole Shear Test.

Triaxial Shear Tests

A determination of R^2 requires paired x and y values, and the Mohr circle representation of triaxial shear test data puts both on the x-axis. A *stress path* can be drawn whereby each Mohr circle is represented by a point at its maximum ordinate, in which case random variability is directed at 45° instead of 90° to the x-axis. A linear regression and calculation of R^2 can be performed by rotating the axes (Handy, 1984).

Problems

1. Why is the friction angle of sand increased by compaction, and how can this effect be demonstrated in a triaxial shear test?
2. Why should dilatancy affect measurements of soil internal friction but not cohesive shear strength?

3. Direct shear tests yield the following results:

Test No.	Normal Stress, kPa (psf)	Shear Stress at Failure, kPa (psf)
1	14.4 (300)	29.0 (695)
2	28.7 (600)	35.7 (745)
3	50.3 (1,050)	49.5 (950)

Use a computer spreadsheet to draw the shear diagram and determine the soil cohesion, angle of internal friction, and R^2.

4. Triaxial compression tests of three test specimens give the following data at failure:

Test No.	Lateral Stress, kPa (psf)	Axial Stress at Failure, kPa (psf)
1	47.9 (1,000)	230 (4,800)
2	95.8 (2,000)	364 (7,600)
3	144 (3,000)	469 (9,800)

Assuming no influence from excess pore water pressure, draw the Mohr diagram and determine c' and ϕ'. Where is random variability, and can it affect R^2?

5. Repeat Problem 4 with the following measured pore water pressures at failure:

Test No.	*u*, kPa (psi)
1	21.4 (3.1)
2	34.5 (5.0)
3	66.2 (9.6)

6. Explain why excess pore water pressure is most likely to occur during initial loading of a foundation. How can it be measured and kept under control?
7. In Fig. 6.7 the first data point is rejected. What is a likely explanation?

References

Handy, R. L., "Linearizing Triaxial Test Failure Envelopes," *ASTM Geotechnical Testing Journal*, 4(4):188–191, 1984.

Lambe, T. W., and Whitman, R. V., *Soil Mechanics*, John Wiley & Sons, New York, 1979.

Further Reading

Das, B. M., *Principles of Foundation Engineering*, 8th ed., Cengage Learning, Boston, MA, 2016.

Vesic, A. C., "Bearing Capacity of Shallow Foundations," In H. F. Winterkorn and H. Y. Fang, eds., *Foundation Engineering Handbook*, Chap. 3, Van Nostrand Reinhold, New York, 1975.

Christian Otto Mohr (1835–1918), German civil engineer and professor who devised the Mohr circle that is a basic tool in geotechnical engineering.

CHAPTER 7

Shallow Foundation Bearing Capacity

A Factor of Safety of 1.0 Indicates a 50 Percent Probability of Failure

7.1. Bearing Capacity versus Settlement

The shift from foundation settlement to bearing capacity failure is like comparing a car crash with a leaky tire. Although settlement slows down as soil consolidates and becomes stronger, shearing involved in a bearing capacity failure will weaken the soil so the rate speeds up. Settlement can be tolerated as long as it is small and evenly distributed among all column and wall foundations, but a bearing capacity failure spells disaster. Therefore, bearing capacity failure must be checked. If for any reason an individual column should settle excessively, floors can collapse and point to the offending column with a "funnel of failure."

Temporary Excess Pore Water Pressure

The most critical time for foundation bearing capacity can be when loading exceeds preconsolidation pressure of a saturated soil and the soil starts to consolidate. As load starts to be transferred to pore water pressure there is no increase in intergranular friction. The result can be as depicted in Fig. 5.8. Storage facilities such as grain bins are especially vulnerable because loads are heavy and must be evenly distributed. A simple way to ensure safe first-time loading is to monitor the soil pore water pressure using *piezometers,* discussed in the next section.

Unanticipated Loading

Sometimes a foundation is subjected to loads that were not included in the original design, particularly in warehouses where loads may be concentrated. Adding an additional floor level to an existing structure can gain space, but the soil must be evaluated and existing foundations measured.

7.2. Fair Warning

Settlement often is monitored both during and after construction. After construction as loading stays constant, the rate of settlement should decrease as in a consolidation test. If it does not decrease, it indicates a viscous behavior called *soil creep.* If creep is not stopped and the settlement rate starts to increase, failure is inevitable and requires immediate evacuation.

Two Kinds of Decrease in Pore Water Pressure

A load-induced increase in pore water pressure should be followed by a slow decrease, as the soil consolidates and water drains out of the soil. However, a *sudden decrease* in pore water pressure will spell trouble because the cause may be dilatant shearing, which means that a bearing capacity failure has started.

Static pore water pressure in soils can be measured with *piezometers*, which are small pipes with filters at the ends so that they can act as wells. Water levels in the piezometers can be measured with a tape that makes electrical contact with the water. The elevation head measured with the tape equals pressure head in the soil, which subtracts from the normal stress and friction. If piezometer readings start rising to dangerous levels, loading is stopped so that there is time for the water to drain away.

Drainage

Excess pore water pressure in saturated, compressible silt and clay foundation soils can be controlled with an array of vertical *sand drains*, borings that penetrate through the consolidating layer and are filled with sand. A layer of sand on top supplemented with tile drains can allow lateral drainage out from under a floor or foundation. Another alternative is to "stitch" in plastic drains with what in effect is a gigantic sewing machine. When failure is imminent, a process used to halt landslides in soils containing expansive clay minerals may be used, and involves pattern borings that extend through the failure zone and are filled with quicklime that immediately reacts with and dries the soil, and then proceeds with a chemical reaction that permanently hardens the soil by increasing the plastic limit. This option is discussed in Chap. 4.

Sensitive Soils

As discussed in Chap. 4, even a small amount of shearing can trigger a failure in sensitive soils; therefore, those soils are avoided for supporting foundations. They do not contain expansive clay and they do not react with lime. Low-density, water-saturated sand can suddenly liquefy into quicksand when given a shake, such as by an earthquake. Preventative treatments are discussed in Chap. 4. Earthquake recurrence intervals do not respect averages.

7.3. Foundations on Compacted Soil Fill

Cut-and-fill operations are the rule rather than the exception on construction sites. Fill soil that is to be used to support roads or foundations is specially processed to ensure stability and an adequate foundation bearing capacity. The soil moisture content is carefully adjusted in order to minimize the effort required to compact it to a suitable density to prevent consolidation and provide adequate shearing strength, that is, bearing

Figure 7.1 Using a scraper to dilute troublesome soil layers for use in compacted fill, suggested by Dwayne McAninch.

capacity. The relationship between dry density and compaction moisture content with a standardized compactive effort was discovered in 1933 by R. R. Proctor, a county engineer for Los Angeles county.

The Proctor density test is routine in geotechnical engineering, and standardized test procedures are described in ASTM D698. The moisture content that gives the highest density is the *optimum moisture content* and is a few percent lower than the saturation moisture content to prevent the development of excess pore water pressure.

Procedure and Performance Tests

Fill soil is spread and compacted in layers, and the compacted soil unit weight and moisture content are routinely measured to ensure compliance with specifications. Nuclear test gauges are commonly used, and it should not be left to a contractor or representative to select the time and test sites.

Cut-and-Fill

The "A horizon" topsoil is valuable as a top dressing, and usually is stripped off and saved for use later to support grass. The "B horizon" subsoil often contains a concentration of clay that can be expansive, but it can be diluted by using a scraper or other machine to cut across layers, as shown in Fig. 7.1. The troublesome soil then is isolated in small pockets so that the moisture content is stabilized.

7.4. Bearing Capacity Equations

The failure illustrated in Fig. 7.2, like many bearing capacity failures, does not require an equation, only common sense. A foundation excavation that removes lateral support from soil under an adjacent foundation is setting up a condition for failure of that foundation. Even excavating down to a foundation level can remove the confining influence from a surcharge load and decrease the factor of safety (FS) against a bearing capacity failure. This is revealed by bearing capacity equations developed by Terzaghi, as it is not only the soil underneath a foundation that supports the foundation, but also the soil alongside that exerts enough pressure to keep the soil underneath from squeezing.

Equation Development

The contributions of W. J. M. Rankine to foundation bearing capacity are discussed in the Introduction. He adapted his solution for soil pressures on retaining walls for foundation bearing capacity (Fig. 7.2), where vertical lines extending down from foundation edges represent imaginary walls. It is a tendency of the soil under a foundation

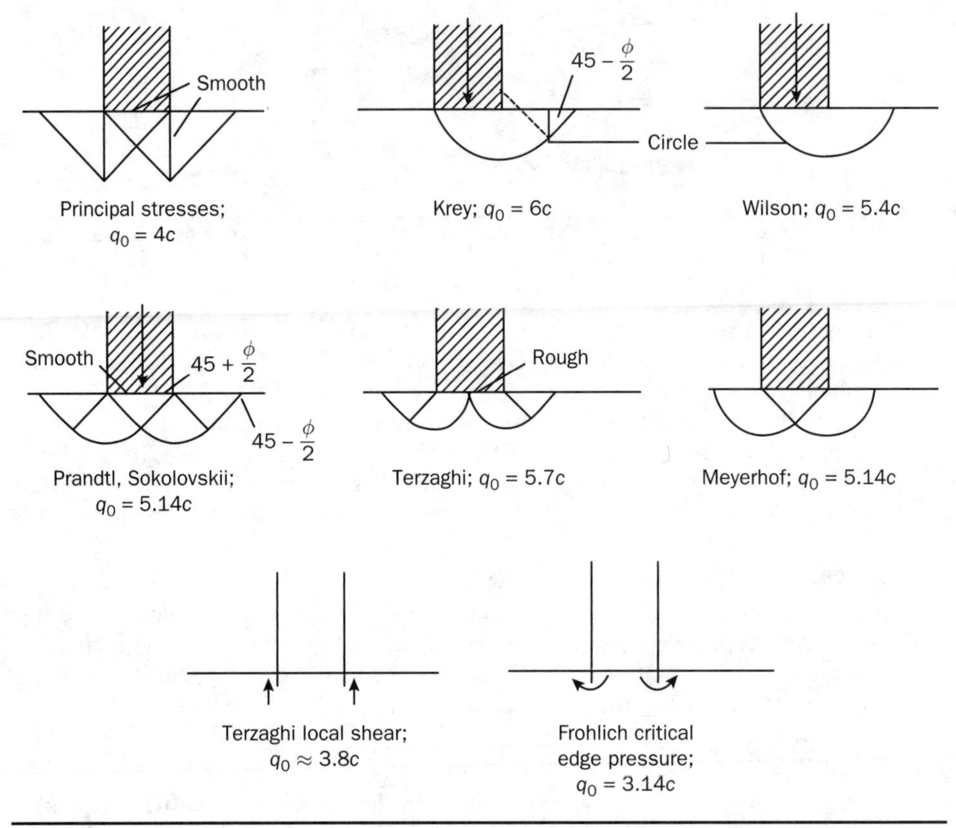

FIGURE 7.2 Some assumed foundation bearing capacity failure geometries and solutions for cohesive soils. (Image source: *Geotechnical Engineering: Soil and Foundations Principles and Practice* by Richard L. Handy and Merlin G. Spangler. © 2007, McGraw-Hill Companies.)

to be pushed outward; therefore, it is in an active (pushing) state against insides of the imaginary walls and creates a passive (being pushed) state on the outside. The concept is seminal and is used with other failure geometries (Fig. 7.2). Geometries involving curved failure surfaces can better explain tilting that is common in bearing capacity failures.

For purposes of comparison the failure geometries shown in the figure are for cohesive soil with no internal friction; therefore, mathematical solutions can be expressed in terms of the soil cohesion. All the failure geometries are two-dimensional, so solutions are for linear foundations. With the exception of the Rankine model the failure geometries shown in the figure provide a range in bearing capacity 5.14 to 6 times the soil cohesion, which is a narrow range considering the variations in failure geometry.

The lower diagrams in Fig. 7.2 show edge or perimeter shear that can be a first step toward a general shear failure such as that shown in Fig. 5.8.

7.5. Prandtl-Terzaghi Analysis

Ludwig Prandtl (1875–1953), a German engineer and physicist, developed a mathematical model for the resistance of a metal surface to penetration by a punch. Terzaghi's

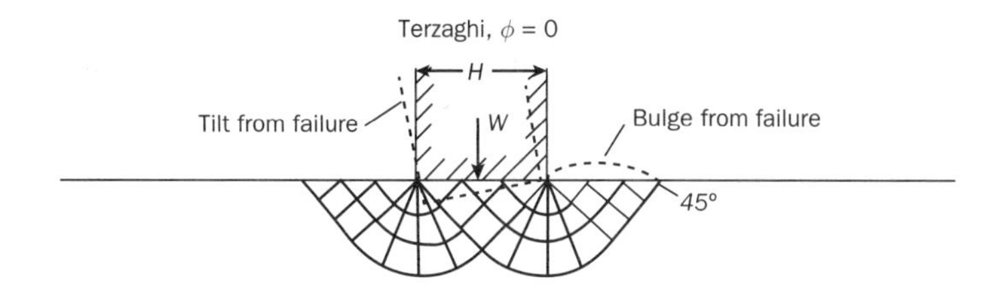

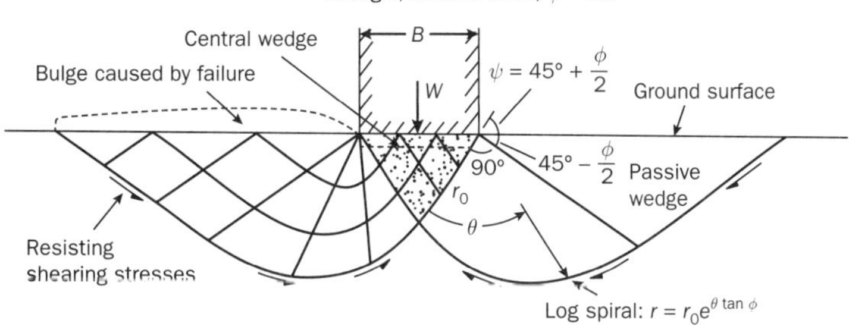

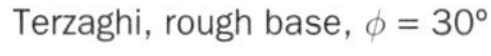

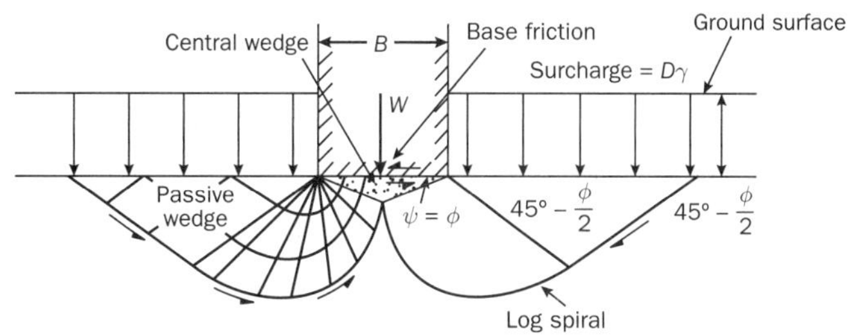

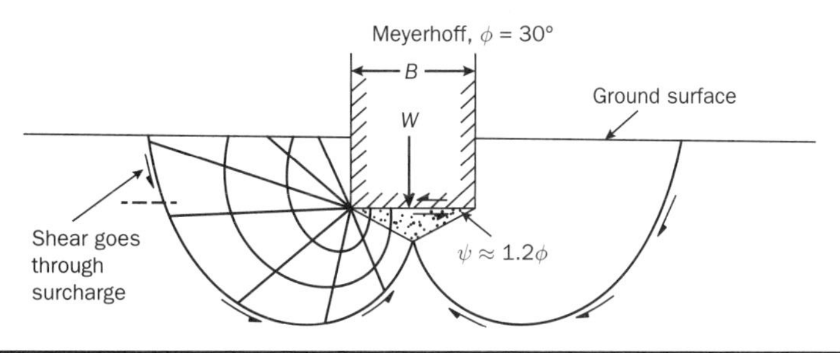

FIGURE 7.3 Terzaghi and Meyerhof approaches to punching shear bearing capacity failures. The top diagram is based on a developed friction angle of zero because of excess pore water pressure. (Image source: *Geotechnical Engineering: Soil and Foundations Principles and Practice* by Richard L. Handy and Merlin G. Spangler. © 2007, McGraw-Hill Companies.)

education initially was in mechanical engineering, and he adapted Prandtl's analysis for calculating foundation bearing capacities. Curved parts of the failure surfaces are *log spirals* (Fig. 7.3). This shape is based on a shear failure angle inclined at $45° \pm \phi/2$ to a normal stress, which is based on Mohr's failure theory that is presented later in this chapter.

In order to simplify the solution, the weight of soil enclosed in the spiral is not included in the analysis, as it will influence both the acting force and the resisting force through its effect on friction. A major part of the resistance is from passive wedges where the weight is included. If ϕ is zero, the log spiral becomes a circle (upper diagram in Fig. 7.3).

Question: Why do the failure geometries shown in Fig. 7.3 include soil resistance on both sides but usually only one side fails?

Answer: Consider a horizontal beam supported by a rope at each end: Both ropes hold the beam up, but as the beam is loaded to failure only one rope breaks.

Rough Base, Smooth Base

Terzaghi solved for two conditions: for a rough foundation where friction will tend to retain soil under a foundation, and a smooth foundation where it can more readily squeeze out. The difference is reflected in *bearing capacity factors* shown in Table 7.1.

Meyerhof's Modification

Prof. G. Geoffrey Meyerhof (1916–2003), at the Technical University of Nova Scotia, solved for a most critical value of basal shear resistance, and as illustrated in the bottom diagram in Fig. 7.3, also included shearing resistance afforded by overburden soil, but Terzaghi considers it to only act as a surcharge weight. The Meyerhof geometry probably is more realistic, but the surcharge soil often will be different or less well compacted than the foundation soil.

7.6. Terzaghi Bearing Capacity Factors

Terzaghi greatly simplified bearing capacity calculations by assigning four separate contributing factors: foundation width (B), soil cohesion (c), foundation depth (D), and the soil angle of internal friction (ϕ), which affects the shape of the log spiral and therefore the other three factors. The first three *bearing capacity factors* are designated by N's:

$$q_o = \frac{\gamma B}{2} N_\gamma + cN_c + \gamma D N_q \tag{7.1}$$

where q_o is *the ultimate bearing capacity* in force per unit area, γ is the soil unit weight, B is the foundation width, c is the soil cohesion, and D is the depth of the foundation. Theoretical N values determined for various friction angles and for rough and smooth foundation bases are listed in Table 7.1.

The three N terms in Eq. (7.1) designate contributions from the foundation width B, soil cohesion c, and foundation depth D. The most abused term may be the third, when one side of a foundation that has been properly designed is uncovered by an adjacent excavation (Fig. 7.4).

Terzaghi Rough Base General				Local Shear		
ϕ	N_γ	N_c	N_q	N_γ	N_c	N_q
0	0.0	5.7	1.0	0.0	5.7	1.0
5	0.5	7.3	1.6	0.3	6.7	1.4
10	1.2	9.6	2.7	0.7	8.0	1.9
15	2.5	13	4.4	1.2	9.6	2.7
20	5.0	18	7.4	2.0	12	3.8
25	9.7	25	13	3.2	14	5.3
30	20	37	22	5.0	18	7.4
35	42	58	41			
40	100	96	81			
45	300	170	170			
Terzaghi Smooth Base General				**Local Shear**		
ϕ	N_γ	N_c	N_q	N_γ	N_c	N_q
0	0.0	5.1	1.0	0.0	5.1	1.0
5	0.5	6.5	1.6	0.3	6.0	1.3
10	1.2	8.3	2.5	0.7	7.0	1.8
15	2.6	11	3.9	1.2	8.3	2.5
20	5.4	15	6.4	2.1	10	3.5
25	11	21	11	3.4	12	4.6
30	22	30	18	5.4	15	6.4
35	48	46	33			
40	110	75	64			
45	270	130	130			

Exponential data fits for programming, all $R^2 > 0.984$:

$Y = ae^{bx}$ where x = friction angle in degrees

Rough Base General			Rough Base Local	
	a	b	a	b
N_γ	0.2337	0.153	0.2089	0.1094
N_c	4.531	0.0749	5.4537	0.0389
N_q	0.8326	0.1136	0.9633	0.0683
Smooth Base General			**Smooth Base Local**	
N_γ	0.2445	0.1532	0.2026	0.1129
N_c	4.0772	0.0714	4.9006	0.0364
N_q	0.8292	0.1075	0.9468	0.0637

Table 7.1 Terzaghi Bearing Capacity Factors for Use in Eq. (7.1)

Figure 7.4 In the foreground is the future site of the "Cave-in-Restaurant," where the foundation excavation extended deeper than the adjacent existing foundation. (Photo courtesy of Prof. J. M. Hoover.)

Example: A sandy lean clay soil has $\phi = 18°$, $c = 700$ lb/ft^2 (33.5 kPa) and $\gamma = 120$ lb/ft^3 (18.8 kN/m^3).

(a) Estimate bearing capacity factors from interpolating data in the table and from the exponential equations for a rough footing surface.

(b) Evaluate the relative influences of footing width and depth on supporting capacity of a footing 3 ft (0.9 m) wide and 4 ft (1.2 m) deep.

(c) With an FS = 3, would removing the depth factor cause the footing to fail?

Answer: (a) From Table 7.1 for a rough footing linear interpolation, $N_\gamma = 4.0$, $N_c = 16$, and $N_q = 6.2$. From the equations $N_\gamma = 3.67$, $N_c = 14.7$, and $N_q = 6.43$. Using the latter values,

$$q_o = \frac{\gamma B}{2} N_\gamma + cN_c + \gamma D N_q$$

$$q_o = \tfrac{1}{2}\,(120 \text{ lb/ft}^3)(3 \text{ ft}) \times 3.67 + (700 \text{ lb/ft}^2) \times 14.7 + (120 \text{ lb/ft}^3)(4 \text{ ft}) \times 6.43$$
$$= 660 + 10{,}290 + 3{,}086 = 14{,}036 \text{ lb/ft}^2 \text{ (14,900 lb/ft}^2 \text{ using interpolated factors)}$$

It will be noted that these are foundation *pressures* and must be multiplied by foundation widths to obtain a supporting capacity in lb/ft or kN/m.

(b) Footing width 660 lb and footing depth 3,086 lb: width 18 percent, depth 82 percent.

(c) Without the depth factor the footing would fail.

Local Shear

As a foundation punches down into soil it is sheared vertically around the perimeter, which can be a particularly important consideration in sensitive clay. Table 7.1 shows bearing capacity factors modified for local shear and appropriate for sensitive clay, but not for quick clay.

Alternative Solutions

Das (2016) lists seven alternative values suggested by different investigators based on various failure geometries and criteria. For example, with $\phi = 30°$, N_γ values vary from 14.44 to 21.27 and average 16.5, compared with 20 and 22 for Terzaghi rough and smooth bases. The Terzaghi factors are most used but obviously a design must incorporate a generous FS that usually is 3.

One explanation for the variability is for an obvious reason, which is the inability to perform full-scale testing. Scale models can be used, but model theory indicates that when gravity is involved, as the scale factor increases so must the gravitational force. Large centrifuges originally used for preparing astronauts have been adapted for soil testing but have limitations. A surface load is not the same as self-weight that accumulates with depth. An imposed downward-directed seepage force might meet that requirement.

7.7. What Is the *Real* Factor of Safety?

A common FS for foundation bearing capacity is 3, but it normally will be physically impossible to increase the load by that amount. It also is unlikely that there will be a 300 percent error in evaluating the soil shear strength or its unit weight. How can an FS be real when it is physically unlikely or impossible? A slope failure can involve a similar geometry with a curved failure surface, but the FS is calculated based on the soil shear strength and unit weight, and the position of a groundwater table and possible seepage forces. The part that is missing is the foundation load. The FS represents a ratio of the total shearing resistance that can be mobilized along a failure surface divided by the sum of the forces tending to cause slipping along the same surface. As the geometry is not known, the problem is indeterminate; therefore, it is estimated through trial-and-error. A similar argument can be made for a bearing capacity failure.

Slope stability is estimated by drawing trial failure surfaces, then summing up acting and resisting stresses at the bases of arbitrary vertical *slices* such as illustrated at the right in Fig. 7.5. (The solution is indeterminate because shearing resistance along vertical surfaces between the slices is not known. In the *ordinary method of slides* it is assumed to be zero.) Different failure surfaces are tested.

In Fig. 7.5 a log spiral shape is assumed, and a solution by the *ordinary method of slices* is shown at the right and gives FS = 1.4. The *Terzaghi smooth base* solution gives FS = 3, higher by more than a factor of 2. That is not an insignificant difference. Both are for the same soil and the same failure geometry, so which is more appropriate, and which is likely to be more comforting to a client? On the other hand, what if a client objects to paying extra in order to achieve an FS of 3?

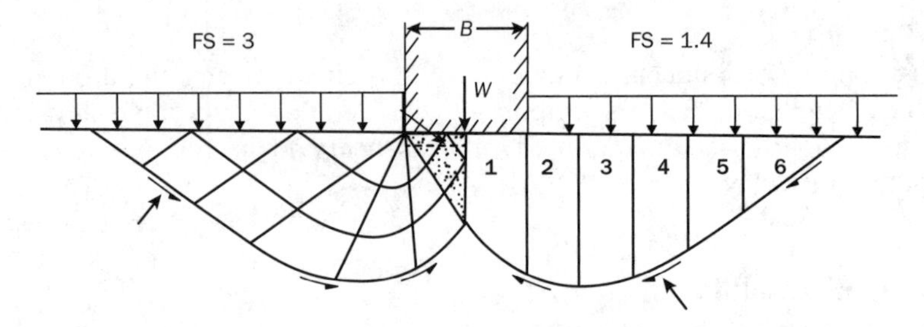

FIGURE 7.5 Two different approaches to an FS, one for foundation loads and the other based on soil strength. Can both be right?

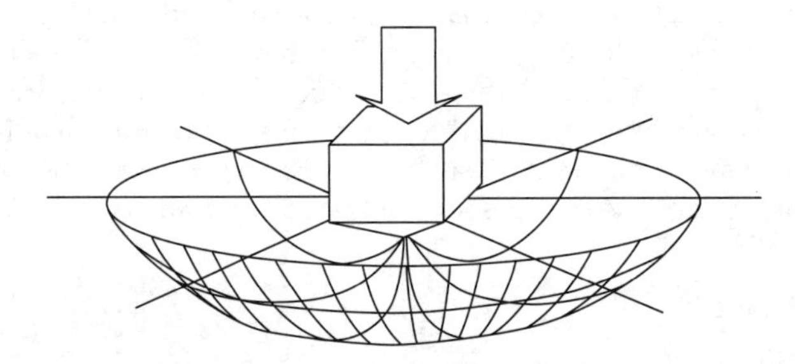

FIGURE 7.6 A square or round footing greatly increases the shear area in the soil, increasing the cohesion component and decreasing contributions of ϕ to the bearing capacity factors. (Image source: *Geotechnical Engineering: Soil and Foundations Principles and Practice* by Richard L. Handy and Merlin G. Spangler. © 2007, McGraw-Hill Companies.)

7.8. Bearing Capacity in 3D

Foundation bearing capacity solutions are two-dimensional, and both analytical and numerical (finite element) solutions struggle with three dimensions. As depicted in Fig. 7.6, a square or round foundation can greatly increase the volume and area involved in shearing compared with a linear foundation. The larger area should increase the role of cohesion, and by spreading out the normal stress should decrease the role of internal friction.

Two-dimensional Terzaghi factors have been modified for three dimensions based on results from physical tests. As expected, N_c is increased and N_γ is decreased. For small footings the smooth-base geometry appears most appropriate. Bearing capacity factors suggested by Prof. Aleksandar Vesic (1924–1982) at Duke University are listed in Table 7.2.

ϕ	N_γ	N_c	N_q
0	0	6.2	1
5	0.27	8.0	1.7
10	0.73	11	2
15	1.6	15	5
20	3.2	21	8.7
25	6.5	31	16
30	13	49	29
35	29	79	57
40	66	140	120
45	160	270	270

$N_\gamma = 0.1414e^{0.1542x}$ $N_c = 4.2666e^{0.0863x}$ $N_q = 0.7835e^{0.125x}$
where x = friction angle in degrees

Table 7.2 Bearing Capacity Factors for Circular or Square Smooth-Base Foundations, Modified After Vesic (1975)

7.9. Eccentric Loading

Off-center or eccentric loading can occur on foundations close to a property line, or with a combined footing that supports both a column load and a wall load.

Eccentricity e is defined as the distance from the central axis of load to the centroid of a bearing area, the maximum value to be $e = 0.5B$, in which case the load is centered over the edge of the footing and the contact area in effect becomes a line.

Prof. Brinch Hansen (1909–1969) at the Technical University of Denmark performed a comprehensive study of bearing capacity factors, and suggested that a footing width can be reduced for use in bearing capacity calculations using the following equation:

$$B' = B - 2e \tag{7.2}$$

Note that if $e = 0.5$, the B' foundation width reduces to zero.

Foundations for Retaining Walls

Eccentric loading is inevitable for retaining walls, as the weight of the wall acts downward and soil pressure acts both laterally and downward, depending in part on friction between the soil and the wall.

Reinforced Earth Walls, also called *mechanically stabilized earth* (MSE), have gained in popularity as they are flexible and easy to design and assemble. They consist of a mass of soil that usually has a width that has a rectangular cross section, is about 0.6 times as

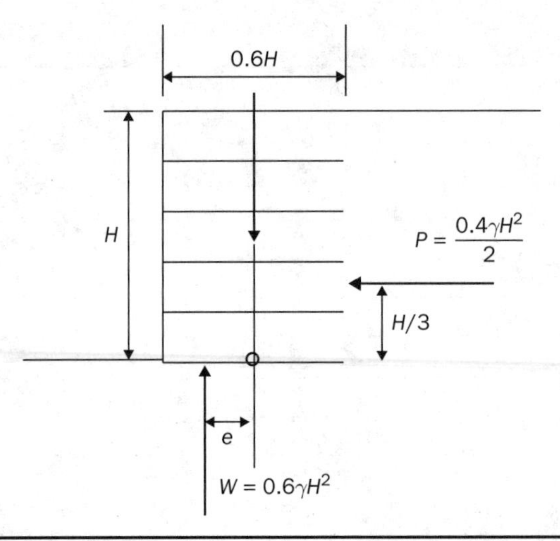

FIGURE 7.7 Example of eccentric foundation loading by an MSE wall. Eccentricity "e" can be further increased when a wall supports the end of a bridge. (Image source: *Geotechnical Engineering: Soil and Foundations Principles and Practice* by Richard L. Handy and Merlin G. Spangler. © 2007, McGraw-Hill Companies.)

wide as it is high, and is internally reinforced with an array of horizontal steel straps. The straps are attached at one end to concrete panels that are the face of the wall. The backfill is compacted sand in order to maximize friction with the strips.

> The use of fibers to stabilize soil dates back to antiquity—straw in the bricks, although the straw also can reduce shrinkage cracking through uniform drying. In the 1960s, French Engineer Henri Vidal observed a stabilizing influence from horizontal layers of pine needles in a pile of beach sand and adapted the idea.

Example: A 10 m (32.8 ft) high MSE wall is to utilize 6 m (19.7 ft) steel straps. The design friction angle of compacted sand behind the wall is 30° and the unit weight is 16.5 kN/m³ (105 pcf). What is the eccentricity of the foundation load and the effective wall thickness?

Answer: Slight lateral yielding of retaining walls mobilizes the soil internal friction to create an active state lateral stress coefficient K_a:

$$K_a = (1 - \sin\phi)/(1 + \sin\phi) = 0.33$$

With reference to Fig. 7.7, lateral force from soil on the wall is

$$P = K_a \times H^2/2 = 0.33(16.5)(10)^2/2 = 272 \text{ kN } (61{,}800 \text{ lb})$$

The weight of the MSE stabilized block of soil is

$$W = 6 \times 10 \times 16.5 = 990 \text{ kN } (223{,}000 \text{ lb})$$

Summing moments around O gives

$$W \times e = P \times (H/3)$$

$$990e = 272\ (10/3) \text{ and } e = 0.92 \text{ m } (3.0 \text{ ft})$$

From Eq. (7.2)

$$B' = 6 - 2(0.92) = 4.16 \text{ m } (13.6 \text{ ft})$$

The reduced width affects the first term in Eq. (7.1), but with sand backfill and no surcharge the other terms in the equation are zero. In this case eccentricity therefore should reduce the bearing capacity for soil support for the wall by 31 percent.

7.10. Mine Collapse

Surveying of underground mines often was not given high priority during early days of mining because of dust, poor lighting, and lease boundaries that tended to be flexible; therefore, it may be no coincidence that mines are called *drifts*.

Shallow Mines

Coal mining often began with tunneling into coal layers exposed along edges of river valleys. Tunnels admitted air and drained away water, and as a result timber supports eventually decayed and tunnel roofs collapsed. Shallow depressions on line at the ground surface are not a good sign. The tunnels still may allow enough air in for spontaneous combustion that emits odors and can help to melt snow, and can be very difficult to extinguish.

Deep Mines

Mine openings that are below the natural groundwater table are likely to be flooded, which preserves timber supports and also reduces the weight of the submerged overlying rock through buoyancy. In addition, roof collapse called *stoping* (with the long "o") fills the tunnel with loose rock, and as it progresses upward the tunnel disappears, to the extent that it can be difficult to locate by drilling. A clue can be a sudden loss of pressure in the drilling fluid. Better is drill cuttings that include saw-cut pieces of wood, rusty nails, or remnants of chewing tobacco tins. Positive evidence usually is required to back up damage claims.

Dangers of Vertical Mineshafts

Vertical shafts were used for mine elevators and air vents, and sometime were only covered with some loose boards when a mine was closed and abandoned. Surface runoff water should be diverted to prevent widening of the opening, and a structural repair can be made with steel rods or mesh supporting rubble fill that is grouted into a solid mass. Areas should be fenced off and heavy loads should be kept off.

Longwall Mining

Room-and-pillar mining can leave about half of the mined material in the ground. A newer procedure called *longwall mining* takes it all. A traveling cutter moves along a face and is followed by hydraulic jacks that lower the mine roof. The overlying rock mass then tends to shear and settle unevenly, so mined-out areas normally are avoided for foundations. Altered drainage patterns and ponding of surface water also can impact agricultural uses.

7.11. A Natural History of Caverns

Limestone caverns develop where infiltrating CO_2-charged water is concentrated and encounters a groundwater table, and mildly acidic H_2CO_3 slowly dissolves the limestone. As nearby stream valleys are cut deeper in geological time the groundwater table is lowered; therefore, cavern levels can be "stacked" with the oldest cavern at the top.

As discussed in Chap. 4, cavernous ground can leave a signature that is readily recognizable from the ground and from the air. It is shallow caverns that have not yet collapsed that are most likely to cause a problem, and can be located with exploration drilling.

7.12. Frost Heave and Footing Depth

Freezing lowers the vapor pressure of water so it freezes and attracts more water. As temperatures stay below the freezing point so that the ground continues to freeze, thin layers of ice accumulate at the freezing boundary and lift and separate the soil to create a condition called *frost heave*. When ice layers in soils under roads thaw they can increase the soil moisture content above the liquid limit, with obvious consequences. Frost heave that lifts a foundation does not have to thaw before creating damage. Prevention normally involves extending foundation below than the maximum seasonal depth of freezing. A guide for design purposes in the United States is shown in Fig. 7.8. A less common option is to separate soil and foundation with an insulating layer such as Styrofoam.

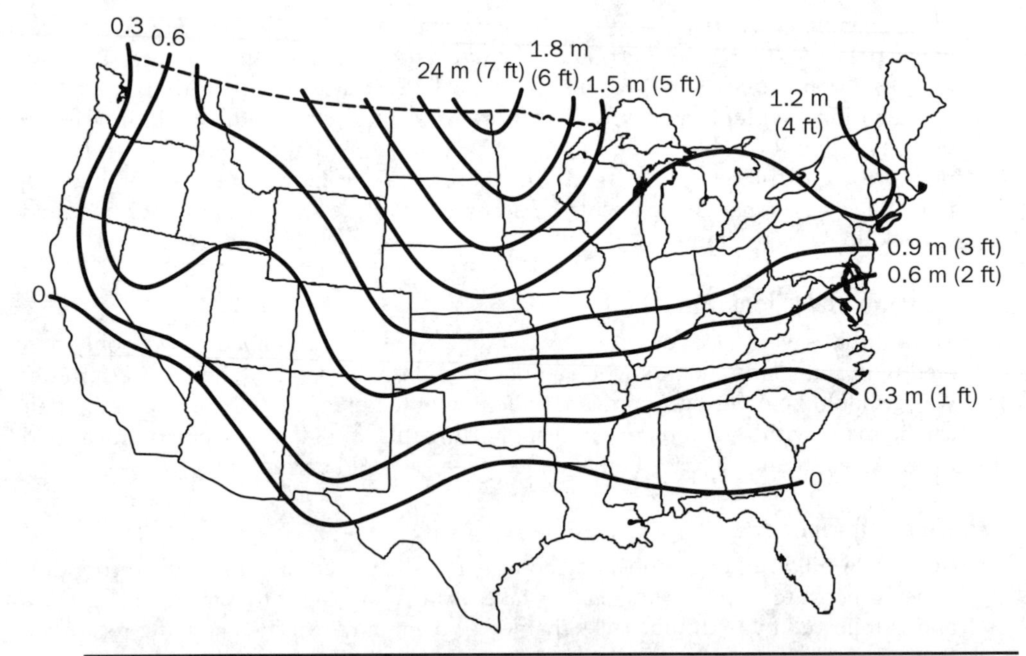

FIGURE 7.8 Guide to maximum depths of soil freezing in the United States. (Image source: *Geotechnical Engineering: Soil and Foundations Principles and Practice* by Richard L. Handy and Merlin G. Spangler. © 2007, McGraw-Hill Companies.)

Arctic Permafrost

In arctic and subarctic areas the mean annual temperature may remain below freezing so the situation is reversed: The lower soil remains frozen and the upper "active layer" thaws during summer months. The underlying permanently frozen soil, or *permafrost,* is a barrier against infiltration of summer meltwater; for example, Arctic areas typically are dotted with lakes and marshes despite having an annual precipitation that would classify it as desert. Permafrost also can be characterized by generous amounts of ice in thin layers, formed and preserved after a paleo-episode of ground freezing.

Polygonal Ground

One of the more striking features of an Arctic plain is an array of hexagonal ridges 10 or more feet across that fit together like floor tiles. They are a result of seasonal drops in temperature that extend down into the permafrost and cause thermal shrinkage cracking. During a subsequent thaw cycle, surface water seeps down into the cracks and freezes to prevent their closing; therefore, seasonal cycling creates "ice wedges" that exert sufficient lateral pressure to push soil up to create high-centered polygons. If the soil is weaker or polygons are larger, they may only push up around the sides to make low-centered polygons. During summers when upper parts of ice wedges melt, polygons are encircled by water. "Beaded streams" are created when streams running across polygonal ground melt into ice wedges. Another consequence is vertical mixing that incorporates organic matter into the soil in a manner similar to that which occurs in expansive clay.

Elongated Lakes

Another indication of permafrost is lakes that would be round but are elongated in the direction of the prevailing wind, as warmer surface water melts the permafrost. When a lake encounters external drainage such as a stream, the lake is drained and the lakebed freezes; thus the cycle begins again. In subarctic timbered areas, lake migration from permafrost melting creates a "drunken forest."

Some Practical Consequences

Heated structures that cause thawing of a permafrost foundation soil will sink into the ground; therefore, the best prevention is to keep the ground frozen. Oil pipelines are supported above the ground on hollow piers that contain Freon which can cycle and keep the ground frozen. A structure that is not heated can create a different problem, by causing thawing to be out of phase with seasonal temperature changes. For example, if a surrounding surface layer of soil starts to freeze while soil underneath the structure remains thawed, the surrounding unfrozen layer becomes a conduit for water and therefore even on a gentle slope it can come up inside the structure and flow out of the windows. A similar natural phenomenon that lifts a frozen soil layer is called a *pingo.*

Methane Release

Permafrost typically contains considerable organic matter; therefore, thawing releases methane that is encapsulated in frozen *clathrate.* Methane is many times more effective than carbon dioxide for contributing to climate change that in Arctic areas is observed to be progressing more rapidly than anticipated. Methane in air slowly oxidizes to form carbon dioxide.

7.13. When Things Go Wrong

Excavation can be required close to an existing foundation and takes away the third, surcharge term in Eq. (7.1). The loss of lateral support for soil under the foundation can be compensated with shoring, as shown at the distant right in Fig. 7.9, with only short sections open at one time. In the figure the shoring was not adequate. Sand was piled against the exposed foundation soil in an attempt to remedy the situation, but it was too late. Near-vertical cracks indicate breaking away of the front wall, and the geotechnical engineer called in to investigate required immediate evacuation of the affected part of the building.

FIGURE 7.9 Inadequate shoring, and an emergency pile of sand that did not prevent settlement and cracking. A hospital waiting room was full of people who were waiting, so they were hustled out of there. (Image source: Author photo.)

Problems

1. A summer home is to be built on a sand beach.
   (a) Estimate bearing capacity and apply an FS for a linear footing 1 m (3 ft) wide on sand at the ground surface. The sand $\phi = 32°$ and $\gamma = 100$ pcf (15.7 kN/m^3).
   (b) Repeat part (a) with the footing at a nominal depth of 1 m.
   (c) Repeat part (b) with the sand temporarily under water.
2. Use data in Problem 1 to estimate an allowable bearing capacity for a square column footing at the ground surface and nominally 0.5 m (1.5 ft) on a side.
3. An owner with a degree in economics sees any word with an S as being spelled with a dollar sign and thinks that an FS of 3 is not realistic, and dumps his concern on the architect, who relays it to the foundation engineer. What is your explanation?
4. What are the assumptions made in the Terzaghi bearing capacity equation? Are they realistic and does it matter?
5. Both wall and column foundations are to be used for a building. Which procedure should be followed?
   (a) Design for equal bearing capacity and check for settlement.
   (b) Design for equal settlement and check bearing capacity.
6. A column settles and therefore floors collapse and create a "funnel of failure." Other columns are investigated, and one shows excessive settlement. What should be an initial focus of the investigation?
7. The bearing capacity of a reinforced earth wall is reduced for eccentric loading. Explain with an example.
8. What is clathrate and why is it important?

References

Das, B. M., *Principles of Foundation Engineering*, 8th ed., Cengage Learning, Boston, MA, 2016.

Vesic, A. C., "Bearing Capacity of Shallow Foundations," In H. F. Winterkorn and H. Y. Fang, eds., *Foundation Engineering Handbook*, Chap. 3, Van Nostrand Reinhold, New York, 1975.

Further Reading

Bowles, J. E., *Foundation Analysis and Design*, McGraw-Hill, New York, 1998.

W. J. M. Rankine (1820–1872), a professor of civil engineering at the University of Glasgow and developer of the Rankine theory of earth pressures.

CHAPTER 8

The Standard Penetration Test in Foundation Engineering

If You Can't Be Delicate Use a Bigger Hammer

8.1. The Empirical Approach

Empirical means experimental, which means to use measurements to establish relationships. However, the first step usually is observational. For example, Hooke's Law for the deflection of a spring might not have happened if he had not observed the deflection of spring. Coulomb's equation for two components of soil shearing resistance was stated in words before it became formalized in an equation. Coulomb's Law, which directs orbits of planets and satellites, began with measurements of tiny forces from static electricity, which must be the ultimate scale factor.

Observations and curiosity can lead to trial explanations that can be tested, and if supported by evidence they can evolve into hypotheses, theories, and laws, or the process can be derailed by fraud and fuzzy thinking. For example, bloodletting was an obvious cure for many diseases because patients who did not die usually got better. The evidence was incontrovertible, and the trouble was that it was the wrong evidence. Once a belief is established, it can become sheltered and difficult to change.

8.2. Soil Penetration Tests

In 1902 a Boston contactor, Col. Charles R. Gow (1872–1949), devised a 1-in. (25.4-mm) diameter drive sampler, standardized the hammer, and used the number of hammer blows to drive the sampler a set distance as a measure of the soil strength. After Gow's firm was purchased by the Raymond Concrete Pile Co., one of his engineers, Harry Mohr (1885–1971), not to be confused with Mohr's circle, redesigned and increased the size of the sampler to the current version that is 2 in. (508 mm) in diameter. It also has a split barrel that is held together by screw fittings at the ends (Fig. 8.1), and a common name is "split spoon."

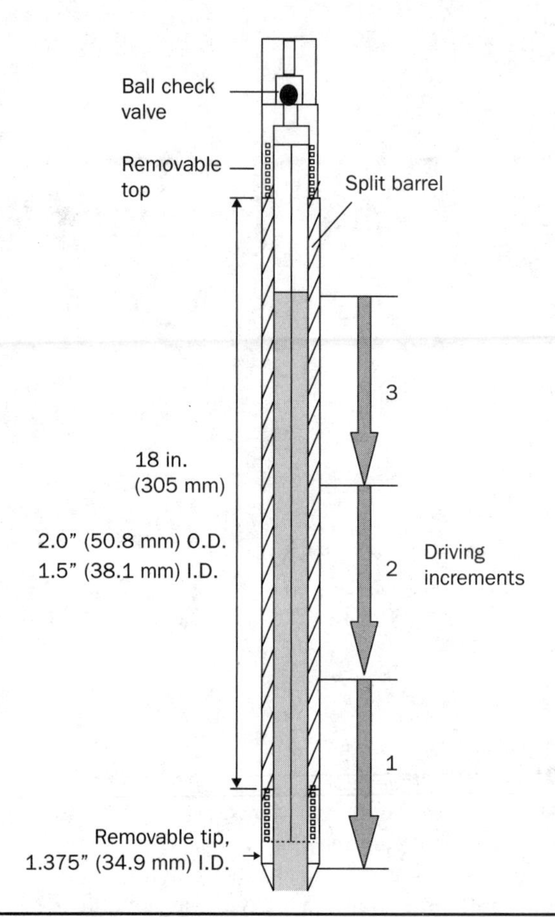

FIGURE 8.1 Standard penetration test ("split spoon") sampler. Blow counts are recorded for driving intervals, then the sampler is removed and ends unscrewed so the long tube halves can be separated to reveal the soil sample.

After the sampler is driven and retrieved it is opened to retrieve a soil sample. The soil is identified and the color described in field notes. One version of the sampler has a removable liner consisting of sections of brass tubes that should not be left out as that will influence the blow count.

In 1947, Terzaghi called the test a *Standard Penetration Test,* and it commonly is designated as the SPT. It was adopted as an ASTM Standard Test Method D1586 in 1958.

> *Test procedure:* A machine boring is made to a test depth, usually with continuous screw-type "flight auger" with a drill bit at the lower end. The auger is removed and replaced with a drill rod and SPT sampler, and the number of blows is counted for three 6 in. (152 mm) advances into soil in the bottom of the boring using a hammer weighing 140 lb (623 N) and a drop distance 30 in. (762 mm). Two of the blow counts, usually for the second and third penetration increments, are added to give "blows per foot" or "bpf." The sampler then is removed and the boring advanced to the next test depth, where the procedure is repeated.

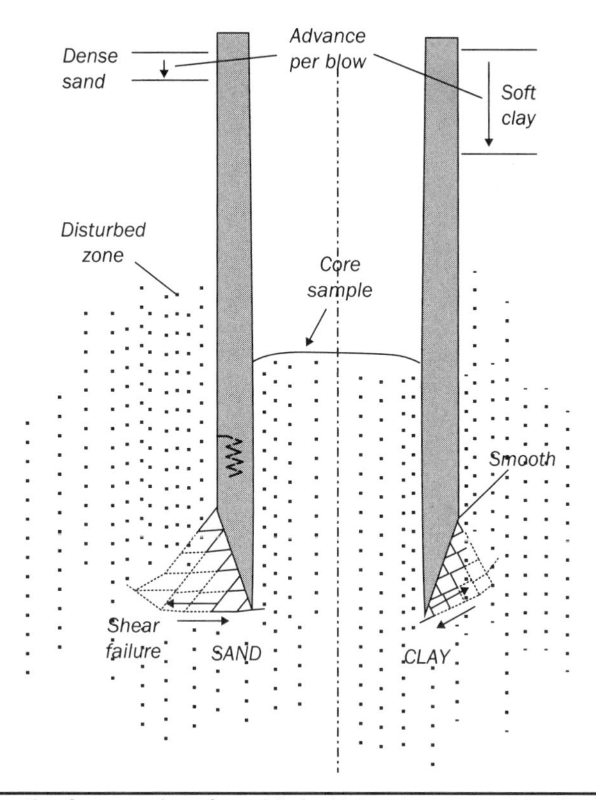

Figure 8.2 Schematic of penetration of sand (left side) and compressible clay (right side) with an SPT "split spoon" sampler.

Selective Test Depths

Continuous testing is not practical, so a common depth interval between tests is 5 ft (1.5 m). That requires that intermediate soil layers must be identified and depths of changes be measured and recorded. A recommended method is to screw helical flight auger down to the next test depth, and then pull it up so that soil layers are preserved between the flights and can be assigned to the correct depths.

Groundwater

Borings normally are left open for 24 hours or longer to allow the groundwater level to stabilize so that it can be measured using a tape with an electrical contact. The boring then is plugged with soil.

Sample Disturbance

Soil immediately ahead of any penetration device is both compressed and remolded. Heavy hammering can increase the pore water pressure ahead of the sampler, and in turn can reduce the penetration resistance and be useful to aid in distinguishing between sand and clay soils.

Initially flat soil layers in the SPT samples are bent downward at edges from friction with inside walls of the sampler, shown schematically in Fig. 8.2. This is indicative of compaction, and soil samples acquired in the SPT are for identification and are not considered reliable for further strength testing, as data can be on the unsafe side.

Several modifications have been made to the basic SPT sampler. The "Modified California Sampler" is 3 in. in diameter and the sample diameter is 2.4 in., compared with the standard diameters of 2 in. and 1.5 in. However, driving a thick-walled sampler still displaces and shears the soil (Fig. 8.2).

The "Pocket Penetrometer"

As a convenience a calibrated hand-held probe called a *pocket penetrometer* sometimes is pushed into soil exposed at the end of the SPT sampler to give a measure of "unconfined compressive strength." The design emphasizes vertical perimeter shear in soil where there is little confining pressure. The results must be properly identified and are not reliable for use in design.

Shelby Tube Samples

Harry Mohr adapted a thin-walled 3-in. (76-mm) diameter steel tube that can be pushed instead of driven into soil and minimize soil displacement and disturbance. The tube was named a "Shelby tube" by the manufacturer. The cutting end of the tube is (or should be) "rolled" inward so the sample is cut slightly undersize. Nevertheless the release from lateral in situ stress may allow the soil to expand and develop friction that accumulates along the length of a tube sample.

After a tube is pushed and withdrawn, ends are sealed to prevent evaporation and it is kept from freezing and shielded from heat of the sun and transported to a laboratory for tests. Soil can be pushed out of the tube from the bottom so that the upper part remains least compressed by internal side friction. Another option is to mill a slot for the length of the tube so that it springs open. Shelby tube samples are appropriate for clay but give no advantage for sand because upon extrusion the samples fall apart.

8.3. SPT in Sand

Excess pore water pressure and cohesion are not factors when testing sands. A correlation between SPT blow counts and sand internal friction suggested by Kishida (1967) is:

$$\phi(\text{degrees}) = (20 \times N)^{0.5} + 15 \tag{8.1}$$

where ϕ is the friction angle and N is the SPT number of blows per foot.

Example: The three incremental blow counts in sand are 12, 16, and 18 blows. Estimate ϕ.

Answer: Adding the last two blow counts gives 16 + 18 = 34 blows per foot

$$\phi = (20 \times 34)^{0.5} + 15 = 41°$$

Depth Correction

Tests conducted near a ground surface in sand can be affected by the ease of lateral displacement, which can reduce lateral stress on the sampler and lower the blow count.

For shallow tests in cohesionless sand (<5 ft or 1.5 m deep), Aggour and Radding (2001) suggest the following correction to ϕ:

$$\phi = (1.5 \times N_{60})^{0.5} + 27° \quad (8.2)$$

Example: Repeat the calculation for a shallow test depth using the previous example and Eq. (8.2).

Answer: $\phi = (1.5 \times 34)^{0.5} + 27 = 34°$, a reduction of 17 percent.

A General Depth Correction

On the basis of empirical tests, Liao and Whitman (1985) suggest a depth correction to the SPT blow count:

$$C_N = (1/\sigma'_v)^{0.5} \quad (8.3)$$

where σ'_v is the effective vertical stress in Tsf, or

$$C_N = (1/100\ \sigma'_v)^{0.5} \quad (8.3a)$$

with σ'_v in kPa.

Depth corrections in Table 8.1 are based on a wet unit weight for sand of 125 pcf (19.6 kN/m³). The "effective depth" is the part of the test depth that is above a groundwater table, plus one-half of the depth that is below the groundwater table, in order

Effective Depth*		
Feet	**Meters**	**Coefficient**
0	0	n.a.
5	1.3	1.79
10	2.5	1.26
15	3.8	1.03
20	5.1	0.89
25	6.4	0.8
30	7.6	0.73
35	8.9	0.68
40	10.2	0.63
45	11.4	0.60
50	12.7	0.57

*Depth above the groundwater table plus one-half of the depth below the groundwater table.

TABLE 8.1 Approximate Depth Corrections to SPT Blow Counts for Sand with a Unit Weight of 125 pcf (19.6 kN/m³). Other Unit Weights Are Used to Calculate σ'_v in Eq. (8.3)

to account for buoyancy. It will be noted that based on those assumptions there is no correction at a depth of 15 ft (3.8 m), but the correction increases substantially below that depth.

Example: The SPT $N = 44$ blows/ft at a depth of 35 ft. The groundwater table is at a depth of 10 ft. What is the corrected blow count?

Answer: The *effective* stress depth is approximated by the depth above the groundwater table plus ½ of the depth below the groundwater table: $10 + (35 - 10)/2 = 22.5$ ft. The depth correction in the table is 0.85. The corrected blow count is $0.85 \times 44 = 37$ blows/ft.

8.4. Soil Mechanics of the SPT

Shearing displacements suggested on the left side of the sampler in Fig. 8.2 are for penetration into sand with $\phi = 30°$. On the right side are shear displacements with $\phi = 0°$, which should apply to saturated clay where all normal stress is carried by pore water pressure. In both cases the zone of shearing expands radially outward, so stress and the amount of disturbance are reduced in accordance with the increasing radius.

As the sampler advances during a test, the contact area between the soil and the sampler barrel increases so that the second and third incremental blow counts usually are higher than the first, and are added to give N in blows per foot. The first count also can be decreased by penetrating soil that has fallen or squeezed into the open boring.

What Might Be Achieved by Subtracting Blow Counts?

A measure of the influence of barrel friction can be obtained by *subtracting* instead of *adding* consecutive SPT blow counts. That is not routinely done, but might give some insight regarding side friction on driven piles. A simple and more accurate determination of side friction suggested by Prof. A. J. Lutenegger of the University of Massachusetts is to measure the torque required to rotate the SPT sampler at the conclusion of each test.

8.5. The SPT Hammers' Biggest Hits

For many years a "doughnut" hammer such as shown in Fig. 8.3 was used to drive the SPT sampler. The hammer is lifted by pulling on a rope so that it tightens on a revolving "cathead." That introduces a personal factor and a resistance factor that can depend on the rope. An enclosed "safety hammer" uses the same lifting arrangement, whereas a "trip hammer" uses an automatic release. As the trip hammer is more efficient, N blow count numbers are substantially lower.

Adjusting the *N* Value

Many thousands of foundation engineering projects have been designed based on correlations with doughnut and safety hammers. Comparing calculated impact energies with measurements indicates that the trip hammer is only about 80 percent efficient because energy loses to heat and inertia of the drill rod string. Examples of measurements and calculated hammer energies at various depths in the same soil are shown in Fig. 8.4.

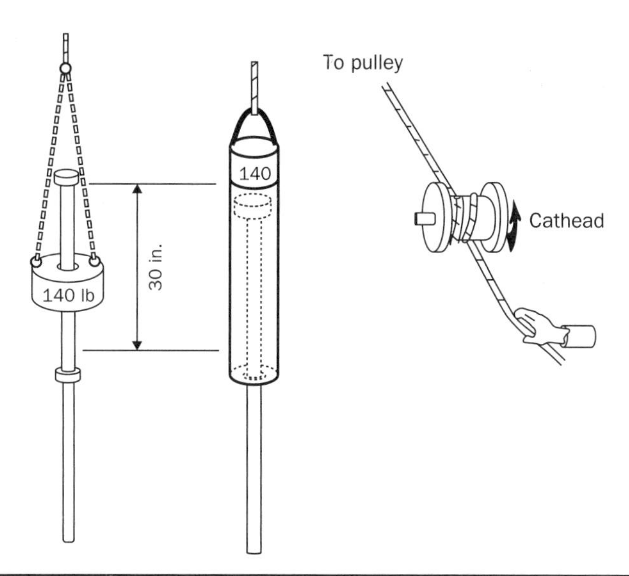

Figure 8.3 (Left) Doughnut hammer, (middle) safety hammer, and (right) cathead and rope release to allow hammer drop. The cathead is considered obsolete for lifting and releasing the SPT hammer. (Image source: *Geotechnical Engineering: Soil and Foundations Principles and Practice* by Richard L. Handy and Merlin G. Spangler. © 2007, McGraw-Hill Companies.)

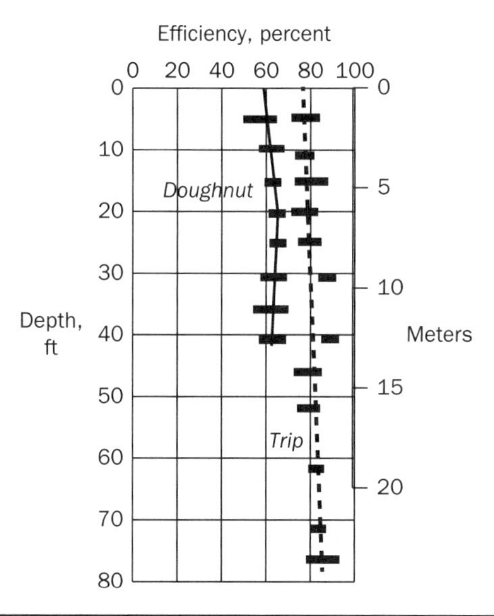

Figure 8.4 Comparable efficiency data for SPT doughnut and automatic hammers. (Adapted from Aggour and Radding, 2001.)

Hammer System	Donut	Safety	Automatic
Average efficiency	63.5%	70.2%	81.4%
Correction factor	1.06	1.17	1.36
Suggested in this book	*1*	*1.2*	*1.4*
Range of published values	0.5–1.0	0.7–1.2	0.8–1.67

Table 8.2 Transferred Energy Efficiency and Correction Factors (modified from Aggour and Radding, 2001)

In order to allow comparisons with older data, blow counts obtained with a trip and automatic hammers usually are numerically converted to equivalent N_{60} blow counts based on energy ratios (ASTM Designation D1586).

Some representative conversion factors are shown in Table 8.2 and may be used if no energy measurements are available. Both measured and corrected values should be included in a geotechnical engineering report. Without the N_{60} correction, N values will be on the safe side for design. Another depth correction that probably is related to an increase in lateral in situ stress is shown by the heavy dashed line in Fig. 8.4, but usually is considered relatively small and on the safe side, and is ignored.

Example: An SPT blow count of 18 bpf is obtained with an automatic trip hammer. What is a suggested value for N_{60}?

Answer: $N_{60} = 18 \times 1.4 = 25$ bpf

8.6. SPT "*N*" Values and Settlement of Foundations on Sand

Predicting the settlement of foundations on sand because of the difficulty of obtaining undisturbed samples for laboratory testing. Figure 8.5 shows a commonly used relationship developed by Terzaghi and Prof. Ralph Peck, University of Illinois. The graph shows corrected (N_{60}) blow counts and an allowable foundation settlement of 1 in. (250 mm) on sand. The horizontal lines at the left reflect a lack of lateral confinement in soil under a narrow foundation.

Lateral confining pressure usually is a function of vertical in-situ stress that can be reduced about one-half by buoyancy if the sand is under water, so the amount of settlement can be approximately doubled by submergence that did not exist at the time of testing.

Example: N_{60} for sand is 25 bpf. What is the allowable bearing capacity for a 2 ft (0.6 m) wide footing at the ground surface?

Answer: According to Fig. 8.5, $q_a = 2.5$ tsf and $Q_a = 5$ T/ft

A Shallow Depth Correction

The lack of lateral confinement at the ground surface should affect the SPT blow count more than a foundation bearing capacity because of its larger contact area. A correction according to Eq. (8.2) may be appropriate: $\phi = (1.5\,N_{60})^{0.5} + 27° = (1.5 \times 25)^{0.5} + 27 = 33°$. Then according to Fig. 8.5, q_a should be about 4.1 tsf instead of 2.5 tsf.

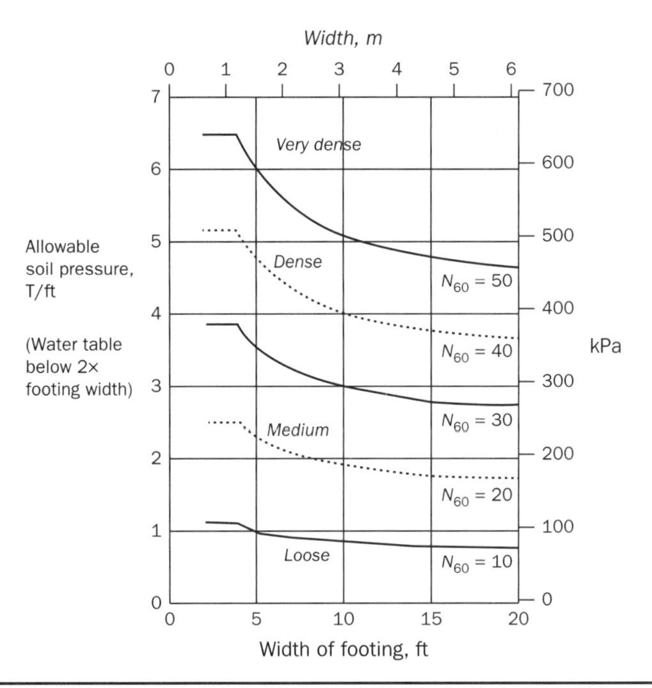

Figure 8.5 Approximate empirical relationships between Standard Penetration Test N_{60} blow counts and 1-in. (25-mm) settlement of a shallow foundation on sand. (Redrawn from Terzaghi and Peck, 1967.)

8.7. Pressure Bulb Correction

Schmertmann et al. (1978) point out a limitation of the Terzaghi—Peck approach because of a pressure bulb effect that is simplified in Fig. 8.6. The relationships approximate elastic theory but are based on screw-plate bearing test data and finite element analyses. The reduction in lateral stress in both x and y directions under a column load, while the decrease under a linear footing is restricted to the x direction, can allow a bearing capacity failure to be more likely to occur under a column than a wall.

Schmertmann et al. suggest the following relationships to predict settlement from cone test data, discussed in the next chapter. The relationships can be converted to SPT blow counts using the following approximations where q_c is in tons/ft^2 and N is in blows/ft.:

$$\text{Sand: } q_c = 4N \tag{8.4a}$$

$$\text{Silty sand: } q_c = 2N \tag{8.4b}$$

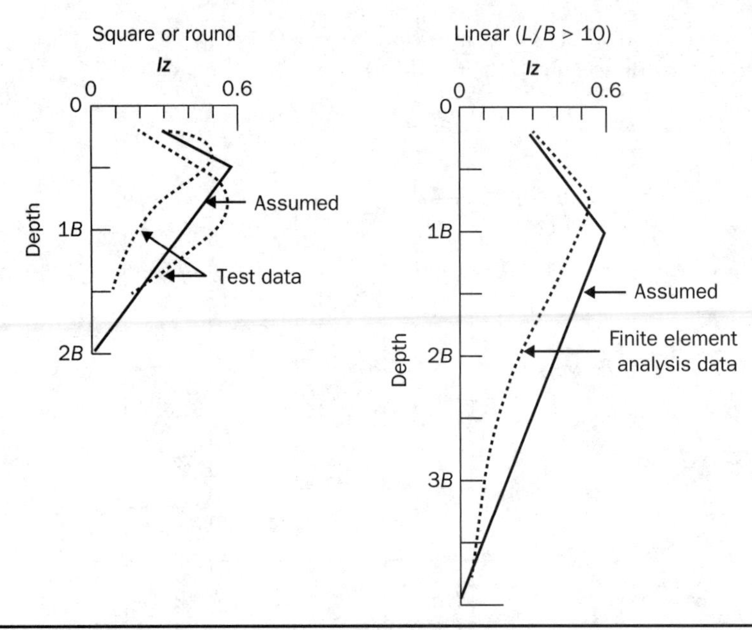

FIGURE. 8.6 Schmertmann et al.'s (1978) suggestion to incorporate influences from pressure bulbs on settlement predictions for settlement of foundations on sand. (Image source: Adapted from *Geotechnical Engineering: Soil and Foundations Principles and Practice* by Richard L. Handy and Merlin G. Spangler. © 2007, McGraw-Hill Companies.)

The equations suggested for settlement are as follows:

$$\text{Column .footings: } S = 0.25Bp/q_c \tag{8.5a}$$

$$\text{Wall footings: } S = 0.31Bp/q_c \tag{8.5b}$$

Example: Estimate the amount of settlement of a 2-ft square foundation carrying 4,000 lb (2 tons) and supported by sand having an average blow count of 30 bpf.

Answer: From Eq. (8.4a), the equivalent cone value is $q_c = 4 \times 30 = 120$ tons/ft². Then from Eq. (8.5a)

$$S = 0.25 \times 2 \text{ ft} \times 2/120 = 0.0083 \text{ ft} = 0.1 \text{ in.}$$

Actual settlement of foundations on sand becomes somewhat ambiguous as it can mainly occur during construction, and the rigidity of box-like structures can aid the transfer of loads away from yielding areas.

The approximations involved in predicting foundation settlement on sand have led to increasing use of in-situ tests such as with the Dilatometer, discussed in the next chapter.

8.8. Bearing Capacity of Sand Based on an Estimated Friction Angle

A foundation obviously must be stable from a bearing capacity failure that involves shearing of the soil. The Terzaghi bearing capacity for a cohesionless soil becomes

$$q_o = \frac{\gamma B}{2} N_\gamma + \gamma D N_q \tag{8.6}$$

where B is the foundation width, D is the depth, and N bearing capacity factors are given in Table 7.1. Submersion under a groundwater table reduces the soil unit weight γ by about one-half.

Example: Testing the previous example for bearing capacity.

Answer: With $\phi = 33°$, from Table 7.1, interpolated relevant bearing capacity factors are

$$N_\gamma = 27.2 \text{ and } N_q = 27.4$$

From Eq. (8.6) if the soil unit weight is assumed to be 100 lb/ft³ (15.7 kN/m³), with $B = 2$ ft (0.6 m)

$$q_o = \frac{100(2)}{2} \times 27.2 + 0 = 2{,}700 \text{ psf}$$

For a footing width of 2 ft, the total load $Q_o = 5{,}400$ lb/ft (80 kN/m). With a factor of safety of 3, the allowable load is $Q_a = 1{,}800$ lb/ft $= 0.9$ T/ft < 2 Tsf., so this is the most critical design criterion. The foundation can be wider and/or deeper.

Example: The footing depth is increased to 3.25 ft (1 m). Then Eq. (8.4) gives

$$q_o = \frac{100(2)}{2} \times 27.2 + 0 + 100(4)(27.4) = 2{,}720 + 11{,}000 = 13{,}700 \text{ psf (656 kPa)}$$

For a footing width of 2 ft, $Q_o = 27{,}320$ lb/ft. With a factor of safety of 3, $Q_a = 9{,}100$ lb/ft (130 kN/m); therefore, the redesigned foundation should be safe from a bearing capacity failure, and the settlement is expected to be less than 1 in. (The number of signficant figures in each N is 3, so answers are rounded to reflect that limitation.)

8.9. Comparisons with Measured Settlements

Terzaghi and Peck (1967) indicated that the correlations shown in Fig. 8.5 probably are conservative, which has been confirmed in later experiments. On the basis of case histories and full-scale load tests, Meyerhoff recommended an increase of 50 percent for bearing capacities indicated in the figure. However, the Terzaghi–Peck relationship continues to be used, especially for projects that do not justify more comprehensive testing and analysis.

Bowles (1988) made additional case history field observations and presented equations for calculating *allowable bearing capacity*, q_a, based on 1 in. of settlement on sand. Those equations have seen widespread use in shallow foundation design for granular soils.

For shallow foundations that are less than or equal to 4 ft in width:

$$q_a = \frac{N_{60}K_d}{2.5} \quad \text{when } B < 4 \text{ ft (1.2 m)} \tag{8.7}$$

where q_a is the allowable bearing capacity in ksf (kPa), N_{60} is the Standard Penetration Test N_{60} value in bpf, $K_d = 1 + 0.33\ (D/B) \leq 1.33$ where B is the foundation width and D is the foundation depth.

When the foundation width is anticipated to be greater than 4 ft,

$$q_a = \frac{N(B+1)^2 K_d}{4B} \quad \text{when } B > 4 \text{ ft (1.2 m)} \tag{8.7a}$$

where B is the foundation width in feet, or 3.3× the width in meters.

A reminder that there are two criteria for a foundation, settlement that involves compression of the foundation soil and a bearing capacity failure that involve shearing.

Case history: Foundations for a retail building bearing on sand were designed with $q_a = 3{,}000$ psf (160 kPa). Six months after construction was completed, the settlement was 0.38 in. (9 mm), less than predicted.

Question: Why? Hint: When may settlement measurements have been initiated?

For more stringent settlement requirements of foundations on sand, q_a can be modified:

$$q_{a(S_t)} = \frac{S_t}{S_o} q_a \tag{8.7b}$$

where q_a is the allowable bearing capacity in ksf (kPa) and S_t/S_o is the arbitrary reduction factor for settlement.

Example: Estimate the allowable soil bearing capacity for shallow foundations for an electrical substation in the western United States with a code frost depth of 3 ft (0.9 m). The soil investigation was performed with Standard Penetration Tests at vertical intervals of 5 ft (1.5 m). They revealed poorly graded fine-to-medium sand with an average $N_{60} = 9$ bpf in the upper 15 ft of the soil profile. The project electrical engineer stated that typical substation footings were 4 ft or less in width.

Answer: Assume $B = 4$ ft (1.2 m); then $D/B = 0.75$. K_d can be solved using Eq. (8.4): $K_d = 1 + 0.33\ (D/B)$ yields $K_d = 1 + 0.33\ (0.75) = 1.25$, which is less than 1.33. Substitution of these values into Eq. (8.4) gives:

$$q_a = \frac{NK_d}{4} = \frac{9 \times 1.25\, q_a}{4} = 2.8 \text{ ksf (149 kPa)}$$

8.10. Foundation Bearing Capacities on Clay Based on SPT or Unconfined Compressive Strength

Foundations on saturated clay can be expected to settle as excess pore water is squeezed out. For flexible structures such as levees where settlement is not critical, the Naval

Facilities Command and the US Army Corps of Engineers relate N_{60} values to the unconfined compressive strength, which is assumed to be undrained:

$$q_u = 300 \text{ pounds per square foot for } 1\ (N_{60}) \text{ blow per foot}$$

$$\text{or } 16 \text{ kPa for 1 blow per } 0.3 \text{ m} \tag{8.8}$$

A layer of empirical correlations can be removed by making direct measurements of the unconfined compressive strength, q_u, on thin-walled tube samples.

Theoretical Foundation Design on Clay Based on Unconfined Compressive Strength

Despite evidence to the contrary it often is assumed that the unconfined compressive strength of saturated clay soil represents a $\phi = 0$ condition. Then according to Table 7.1, $N_\gamma = 0$ and $N_q = 1$ for both the rough and smooth base geometries. The Terzaghi bearing capacity Eq. (7.1) then reduces to

$$q_o = N_c c + \gamma D \tag{8.9}$$

If $\phi = 0$, according to a Mohr circle in Fig. 6.8, the unconfined compressive strength $q_u = 2c$. Substituting this relationship for c gives:

$$q_o = \frac{N_c q_u}{2} + \gamma D \tag{8.10}$$

As specimens in the unconfined test do not leak water, the measured strength includes some contribution from internal friction that will add additional restraint to the equation so it should be on the safe side.

Net Bearing Pressure

A *net bearing pressure* equals q_o minus the overburden pressure at the base of the foundation. That removes the last term in Eq. (8.9). Applying the standard of 3.0 for the factor of safety gives an *allowable soil bearing pressure*, q_a:

$$q_{a(\text{net})} = \frac{N_c q_u}{6} \tag{8.11}$$

Figure 8.7 shows solutions for this equation by Peck et al. (1974) based on unconfined compressive strength with a presumed $\phi = 0$, when depth D and the footing width B are known. This approach does not directly address settlement. However, the lack of lateral restraint in an unconfined compression test should allow the amount of compression to be larger than that which will occur in the field. Settlement can be more clearly defined using procedures outlined in Chap. 6.

Reducing Settlement with a Mat Foundation

Differential settlement can be prevented with a continuous mat (or raft) foundation that is reinforced to bridge over soft spots, and a bearing capacity failure becomes unlikely because of the large width factor (but not impossible, as shown in Fig. 5.8). However, a mat is considerably more costly than individual column and/or wall footings.

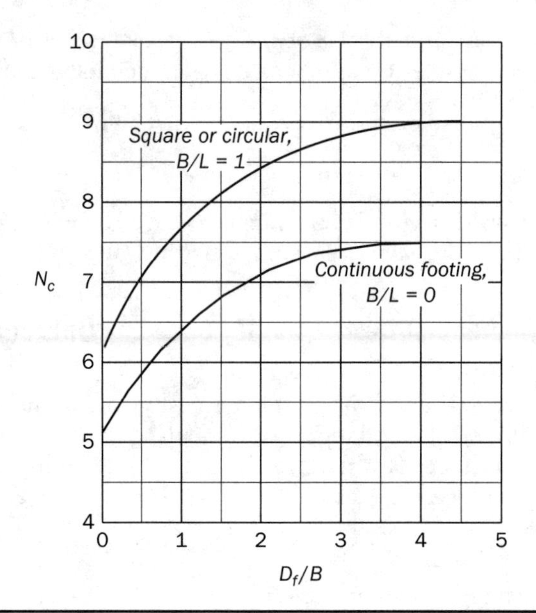

Figure 8.7 Bearing capacity factor N_c *for* square or linear footings on clay soil with assumed $\phi = 0$. (Redrawn from Peck et al., 1974.)

Summary

As indicated above, there are many options for calculating foundation bearing capacity and settlement from SPT N_{60} blow counts, and the selection of a method often will depend on judgment and local experience.

Example: Estimate the allowable soil bearing capacity for a wall footing bearing on lean clay that exhibits an unconfined compressive strength of 1,850 lb/ft². There is a code-mandated frost depth of 3½ ft. The project structural engineer has indicated that a typical footing width for this type of structure is 2 ft.

Answer: The trial ratio $D/B = 3½$ ft/2 ft $= 1.75$. From Fig. 8.7 and the continuous footing curve, the bearing capacity factor $N_c = 7$. Substituting these values into Eq. (8.9) yields:

$$q_a = \frac{N_c q_u}{6} = \frac{(7)(1,850)}{6} = 2,160 \text{ lb/ft}^2 (115 \text{ pKa})$$

which includes a factor of safety of 3.

Problems

8.1. In the SPT, what percent of the cross-sectional area of the sampler is occupied by the soil sample and what percentage by the steel wall of the sampler? In a California sampler? In a 3-in. diameter Shelby tube? What can be done to a Shelby tube to decrease friction between the soil sample and the inside wall of the tube?

8.2. What is the allowable soil bearing capacity for a 5-ft wide column footing bearing on silty lean clay located in the southeastern United States with a code frost depth of 1½ ft. The field exploration revealed the silty lean clay exhibited a median Standard Penetration Test $N_{60} = 7$ bpf.

Answer: 2,300 lb/ft^2 (125 kPa)

8.3. What are some of the problems associated with predicting the settlement of foundations on sand?

8.4. Failure of a column foundation can cause floors to define a "cone of failure." Although column failures are extremely rare, why are they more likely to occur than a wall failure?

8.5. Field SPT blow counts are determined using a trip hammer, whereas empirical correlations were based on experiments using "donut" hammers so blow counts were higher. What can be done to compensate for the discrepancies?

References

Aggour, M. S., and Radding, W. R., *Standard Penetration Test (SPT) Correction.* Final Report SP007B48, Maryland DOT State Highway Administration, 2001. (Available online.)

Bowles, J. E., *Foundation Analysis and Design*, 4th ed., McGraw-Hill, New York, 1988.

Kishida, H., "Ultimate Bearing Capacity of Piles Driven into Loose Sand," *Soils and Foundations*, 73(3):20–29, 1967.

Liao, S. S., and Whitman, R. V., "Overburden Correction Factors for SPT in Sand," *Journal of Geotechnical Engineering*, 112(3):373–377, 1986.

Peck, R. B., Hanson, W. H., and Thornburn, T. H., *Foundation Engineering*, 2nd ed., John Wiley & Sons, New York, 1974.

Schmertmann, J. H., Hartman, J. P., and Brown, P. R., "Improved Strain Influence Diagrams," *Journal of Soil Mechanics and Foundation Engineering*, 96(SM3): 1011–1043, 1978.

Terzaghi, K., and Peck, R., *Soil Mechanics in Engineering Practice*, 2nd ed. John Wiley & Sons, New York, 1967.

Further Reading

Meyerhoff, G. G., "Some Recent Research in the Bearing Capacity of Foundations," *Canadian Geotechnical Journal*, 1(1): 16–26, 1963.

NAVFAC DM-7-2, *Soil Mechanics Design Manual 7.1*, Department of Navy, Naval Facilities Command, Alexandria, VA. 1982. (Available online.)

Rogers, J. D., "Gow, Mohr, Terzaghi, and the Origins of the Standard Penetration Test," Presented at a Joint Meeting of the Association of Environmental & Engineering Geologists, American Society of Civil Engineers, Chicago, IL, 2009. (Available online).

Ralph Peck (1912–2008), eminent geotechnical engineer and consultant, and professor at the University of Illinois.

CHAPTER 9

Probing with Cone Penetration Tests and the Marchetti Dilatometer

Fast and Convenient

9.1. A Classical Approach

A common approach to invention is to try something and see if it works, and if it doesn't work try something else. Insects probably stumbled countless times before they learned how to fly, and reptiles repeated the process and then passed it on to birds. The Wright brothers used a similar approach but also were able to analyze how it works.

In England, John "Blind Jack" Metcalf (Fig. 9.1) evaluated the strength of soils for roads by probing with his staff, and even without eyesight discovered the importance of drainage.

In 1930, soil probing was improved with the development of the "Dutch cone" at the Department of Public Works, the Netherlands. The force required to push a conical tip down into soil and is followed by measuring a separate force required to push a follow-up "friction sleeve" that has the same diameter. The cone is pushed ahead with a slightly smaller "mantle" (Fig. 9.2). Pushing is accomplished with two separate rods.

Figure 9.1 Road engineer John "Blind Jack" Metcalf (1717–1810) and an early tool that was used to determine the strength of soils for roads.

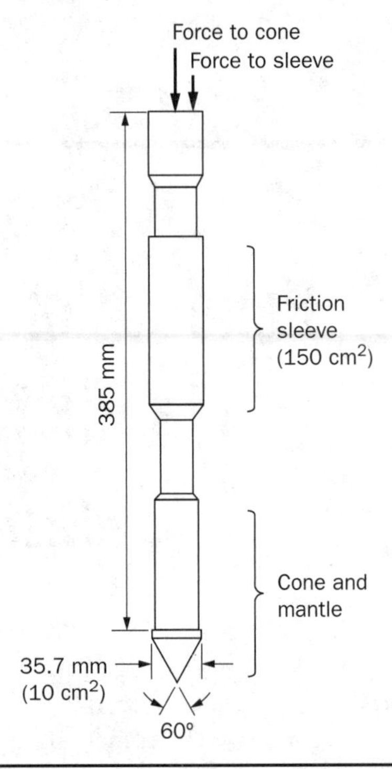

FIGURE 9.2 The mechanical Dutch cone separates forces required to push the cone tip and the follow-up "friction sleeve," giving two test values. (Image source: *Geotechnical Engineering: Soil and Foundations Principles and Practice* by Richard L. Handy and Merlin G. Spangler. © 2007, McGraw-Hill Companies.)

9.2. Pushing versus Driving

Hammering is more efficient than pushing, as anybody can testify who has tried to push a nail. Cones therefore displace less soil than an SPT sampler, and do not acquire a sample. Cones are standardized to have a cross-sectional area of either 10 or 15 cm^2, with respective tip diameters of 35.7 mm (1.40 in.) and 43.7 mm (1.72 in.). The cone tip angle is standardized at 60°, which according to Mohr theory is the failure angle in soil that has an angle of internal friction of 30°.

Displacement of soil by the cone creates a dilatant component to shearing that can help to distinguish between sands from clays. This is shown by empirical correlations along the y-axis in Fig. 9.3. Note the logarithmic scale. Such correlations are important because no samples are obtained but it should be emphasized that the boundaries overlap and are approximate. Computer printouts that are based on discrete boundaries can show discrete layers instead of variations that can be expected to occur within a layer.

9.3. A "Friction Ratio"

Tip resistance in effect represents a miniaturized bearing capacity failure. Sleeve resistance can measure sliding friction, but internal friction in soil also is influenced by

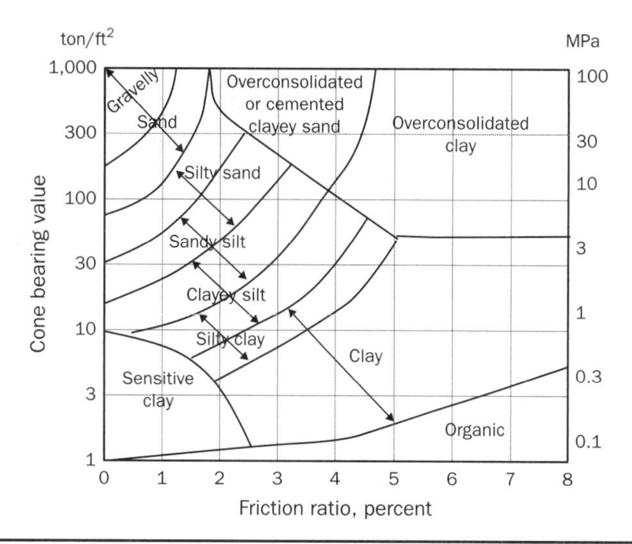

Figure 9.3 Overlapping ranges in soil identifications from cone test data. (The ranges are not shown in ordinary computer printouts.) (Modified after Robertson and Campanella, 1983.) (Image source: *geotechnical Engineering: Soil and Foundations Principles and Practice* by Richard L. Handy and Merlin G. Spangler. © 2007, McGraw-Hill Companies.)

dilatancy, the tendency for soil to expand when it shears as soil grains ride up and over one another. The dilatant contribution to friction is not measured with sleeve resistance, and it also can be influenced by soil *cohesion* (more properly called *cohesive shear strength*).

As shown from left to right in Fig. 9.3, the so-called *friction ratio* typically is higher for clay than for sand! The initial confusion is understandable, but perhaps by the 100th anniversary of the test the ratio can be renamed a *sleeve resistance ratio* or an equivalent.

9.4. Mechanical versus Electrical Cones

Mechanical cones, as shown in Fig. 9.2, can be pushed with conventional truck-mounted drilling equipment, but screw anchors may be required to hold down the back of the truck. Pushing goes by steps, with the distance for each advance of the cone being determined by the length of the mantle.

Electrical cones also have two parts, but pushing can be continuous as the two resistances are continuously measured, and the data stream is recorded and can be graphed as a test is being conducted. Sensitive electronic equipment usually is enclosed in a vehicle dedicated to that purpose, and a suitable reaction base is obtained by jacking from the center of gravity of the vehicle. Stop-and-go still is required to add more push rods, and care is required when feeding and recovering electrical lines.

A representative graph of data is shown in Fig. 9.4. As the sleeve length contacts a much larger range in depths that are affected by advance of the cone tip, the "friction ratio" tends to be averaged.

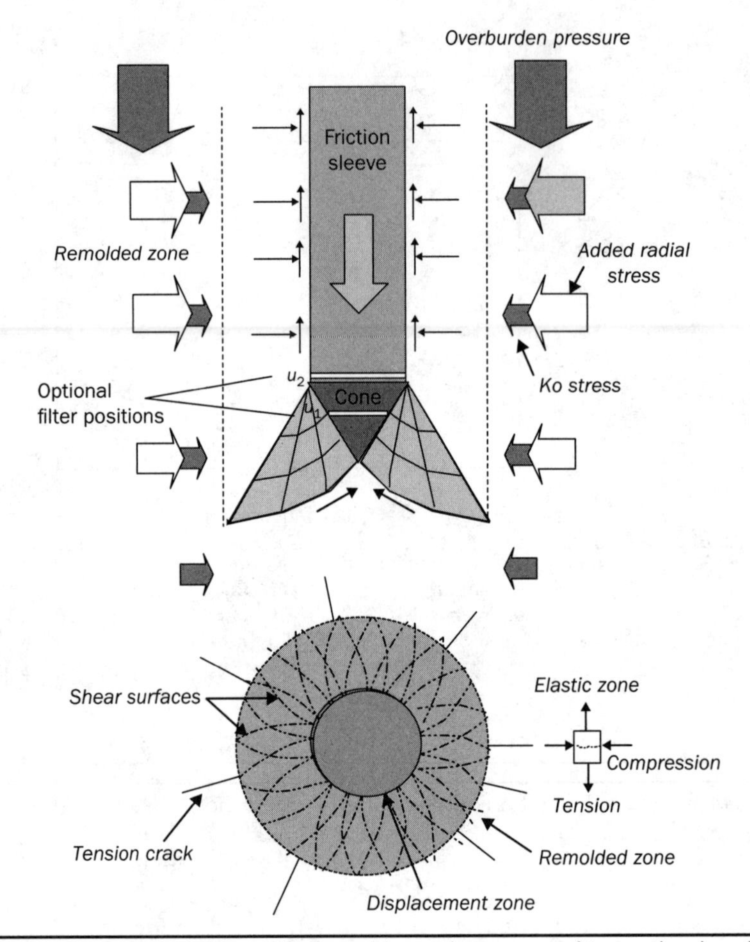

FIGURE 9.4 Soil displacements inferred from pushing a piezocone, and two optional positions for filter positions to monitor pore water pressure.

The Piezocone

Electric cones often incorporate a porous element near the tip for simultaneous monitoring of pore water pressure, which represents a combination of static pressure from a groundwater table and excess pressure developed from compression of saturated soil near the cone.

Two common positions for the porous filter are "u_1" on the cone face and "u_2" at the upper end (Fig. 9.4). Pore water pressure tends to be higher at the u_1 position because the soil is more compressed, but the filter also is more likely to become plugged, and the u_2 position is most commonly used.

Pore water pressure also is sensitive to dilatant or volume expansion of soil upon shearing, so a negative pressure at shallow depths can help to distinguish between saturated dense sand that dilates, and loose sand that compresses. However, dilatant expansion also occurs in expansive clay that has a hard, blocky structure. A blocky soil structure is common in subsoil "B horizons" that also can occur at depth when buried under younger sediments, as on river floodplains.

Decision Time: What Are Advantages/Disadvantages of Cone and SPT?

The SPT requires little in accessory equipment, and yields samples for positive identifications. However, it is many times slower than cone tests, as after each episode of driving, the "split spoon" sampler must be removed and substituted with a drill or auger to advance the boring through the tested zone and down to the next test depth. Penetration tests routinely are only conducted at 5 ft (or an equivalent) depth increment. Soil samples obtained by driving are adequate for identifications but not for strength tests, so SPT sampling often is alternated with thin-walled *Shelby tube* sampling for laboratory strength determinations. The hole left by a tube sampling also can be tested for a rapid determination of soil cohesion and internal friction using a Borehole Shear Test.

If a helical auger is used to advance a boring between tests, it can be screwed down into the soil and then pulled up so that soil is retained between the flights and can be identified. Particularly important can be soil color, as reducing conditions under a groundwater table contribute to a gray color, and oxidation above a groundwater table turns it tan or brown. A mottled mixture of gray and brown can be particularly significant, as it can reveal seasonal high levels of groundwater that can affect foundation bearing capacity and basement drainage. Close to a natural gas leak it can reveal a tinge of green or blue. Drillers should be informed of the relevance of soil color. Color is not revealed in cone tests, but the existing position of a groundwater table can be revealed from piezometer readings.

SPT borings often are left open for a day or more to allow measuring the level of the groundwater table. The SPT has a long history, and still is widely used.

Advantages and Disadvantages of Cone Tests

Cone tests are faster than SPT, but do not yield soil samples, so soil identifications are based on empirical correlations that are approximate. Electrical cone tests are increasing in popularity as the data stream is continuous and can be recorded, graphed, and interpreted on-the-go. Data and approximate soil identifications based on empirical correlations can be printed in color and can be a readily obtained and impressive addition in geotechnical reports. However, the interpretation may be oversimplified and often will distinguish between more soil layers than actually exist, particularly at shallow depths in a weathered soil profile.

Piezocone and Groundwater Table

Pushing of a piezocone may be stopped long enough for the pore water pressure to equilibrate, and the rate of decline also can give an indication of the soil hydraulic conductivity.

9.5. Fracking (Hydraulic Fracturing)

In the 1940s, grout that was pumped under pressure into shallow rock emerged on the other side of a hill and flooded a farmer's field. This led to experiments using water, where pumping pressure was increased and then suddenly dropped when rock layers separated. Sand then was added to the water to keep layers propped open so they could release oil. Electronically controlled directional drilling was added to allow a horizontal array of oil wells to extending outward from a single vertical well like spokes in a wheel. That opened shale deposits for petroleum production, and shale is easier to drill than hard rock.

Fracking produces large amounts of polluted water, and an obvious place to put it is back into the ground by pumping it down into abandoned oil wells. That can produce an unfortunate side effect—an increase in pore water pressure and reduction in effective stress, which can lead to stick-slip along existing faults to produce earthquakes. Stress that is relieved in one place is transferred to another, and in the United States, Oklahoma now leads California in frequency of earthquakes.

9.6. Example of Cone Test Data

Figure 9.5 is an example of piezocone test data from river floodplain soil. Pore water pressure is a signal property, as it tends to be retained in silt and clay layers and not in sand. Pore water pressures measured in the sand are depicted by the heavy dashed line in the right graph and are consistent with a water level at the ground surface, as may be expected on a recently inundated floodplain. The sleeve resistance in the center graph usually is lower than the tip resistance in the left graph; therefore, as previously noted, the "friction ratio" is lower in sand than in clay, which at best is confusing and at worst is more confusing.

The most likely cause of a small negative pore water pressure in the upper 5 m of soil is a clay structure described by soil scientists as "subangular blocky" and indicative

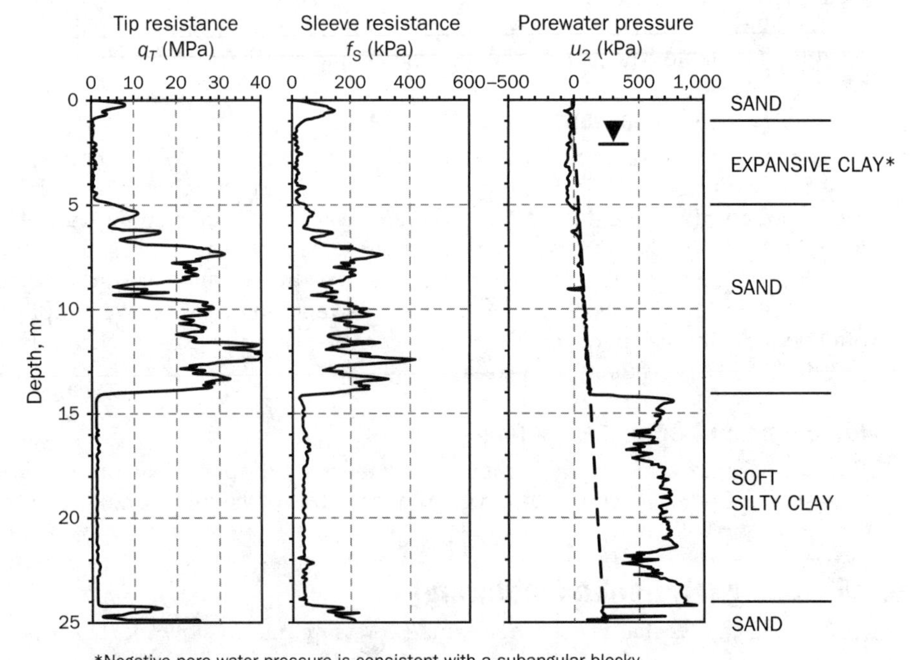

Figure 9.5 Representative piezocone test data from a site on the Mississippi River floodplain. (Adapted from Mayne, 2007.)

of expansive clay. The large thickness suggests repeated intervals of deposition and drying, as commonly occurs on meandering river floodplains. Foundations should extend below the clay and bear on or in the underling sand.

9.7. Normalizing Cone Test Data for Test Depth

Volumetric displacement of soil is 100 percent for the cone but less than 50 percent for the SPT because it cuts out a soil sample, and the disturbance is distributed over a larger perimeter area.

A linear relationship between cone tip resistance and depth can automatically be tested by dividing the tip resistance by the test depth, a process referred to as *normalizing*. Figure 9.6 shows cone data treated in this way, and reveals an unusually close relationship to sand internal friction angles measured in laboratory triaxial shear tests, as excess pore water pressure in sand quickly drains away. Friction angles from cone test data can be used in combination with soil unit weight and with Terzaghi bearing capacity factors to estimate the foundation bearing capacity, but it is important that soil unit weights be corrected for buoyancy for soil under a groundwater table.

Dealing with Dimensions

The formula for sand friction angle in Fig. 9.6 includes a square root term. For the equation to be dimensionally consistent, both the cone tip stress q_t and the effective overburden stress σ_{bo}' are divided by an arbitrary reference pressure to become dimensionless. A convenient reference is atmospheric pressure, 1 bar. The following approximations are useful for making calculations using this graph: 1 bar $\approx$ 100 kPa $\approx$ 1 Tsf.

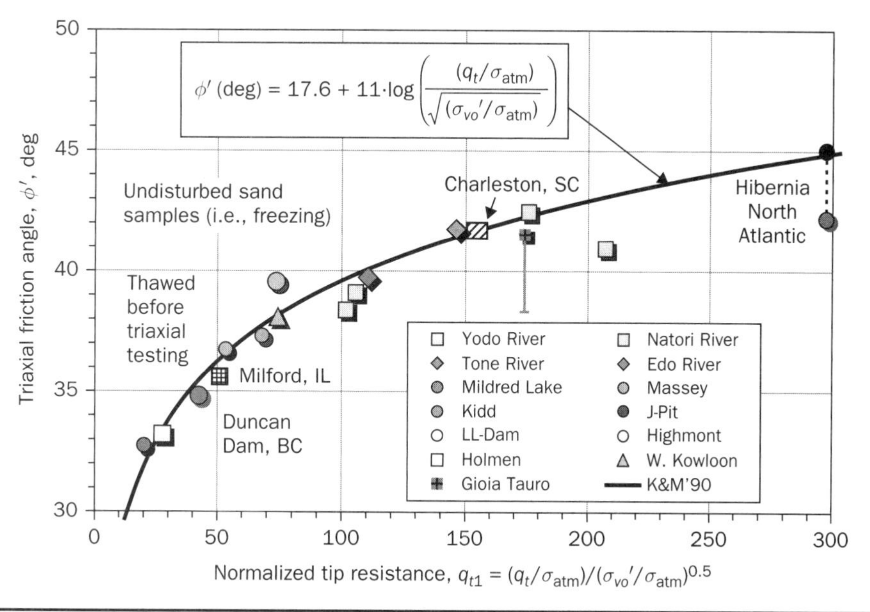

FIGURE 9.6 Friction angles in sand in relation to cone tip resistance normalized for test depth σ_{vo}'. (From Mayne and Niazi, 2017.)

Example: An average cone tip resistance is 2 MPa = 2,000 kPa = 20 atm, in sand 1 m below an anticipated foundation depth of 1 m. Estimate the friction angle.

Answer: Assume a soil unit weight of 100 pcf, which at a depth of 6 ft (or 2 m) gives $\sigma_{vo}' = 600$ psf = 0.30 Tsf = 0.3 atm. Then $q_{t1} = (q_t/1\ \text{atm})/(\sigma_{vo}'/1\ \text{atm})^{0.5} = (230/1\ \text{atm})/(0.3/1\ \text{atm})^{0.5} = 36$ MPa, and ϕ is approximately 35°.

9.8. Cone Test Data and Settlement of Foundations on Sand

Settlement is more likely to be a controlling factor than bearing capacity for foundations on sand. Schmertmann et al. (1978) approximate the pressure bulb with a triangular stress distribution that extends to a depth of $2B$ under square or round foundations, and $4B$ under linear foundations with a length/width ratio is greater than 10. The relationships are shown in Fig. 8.6. The equations are:

$$\text{Square or round: } S = 0.25Bp/q_c \tag{8.5a}$$

$$\text{Linear: } S = 0.31Bp/q_c \tag{8.5b}$$

where B is the footing width, p is the net bearing pressure (bearing pressure minus overburden pressure on an effective stress basis), and q_c is the cone bearing value.

Note that units of p and q_c must be the same, and units for S are the same as for B.

Example: Estimate settlement for a 2-ft wide linear footing exerting a bearing pressure of 800 psf, and cone data from the previous example, $q_c = 2{,}000$ kPa.

Step 1. Convert p into the same units as q_c: $p = 800$ psf = 0.4 Tsf = 40 kPa.

Step 2. $S = 0.31Bp/q = 0.31$ (2 ft)(40 kPa/2,000 kPa) = .012 ft = 0.12 in. = 4 mm (held to one significant figure).

9.9. Cone Tests and Foundations on Saturated, Compressible Clay

An early use of cone tests was to consider the cone as a model foundation, but the model is flawed because tangential tension that has a small influence on a foundation can significantly affect cone resistance. As a practical matter settlement is more likely to be a controlling factor than foundation bearing capacity, but there have been exceptions.

A worst condition for bearing capacity can be based on an assumed "$\phi = 0$" condition. The Terzaghi bearing capacity equation for a $\phi = 0$ soil is

$$q_a = cN_c \tag{9.1}$$

where c is the soil undrained shear strength or cohesion and N_c is the two-dimensional bearing capacity factor. The factor is modified for the cone because of the three-dimensional aspect, fixed cone angle, and smooth contact surface with the soil:

$$q_t = cN_{kt} \tag{9.2}$$

where q_t is the cone tip resistance and N_{kt} is a *cone tip bearing capacity factor*.

Equation (9.2) can be solved for cohesion c, also referred to as the *undrained shear strength*, s_u. However, experimental data indicate a broad range in values of that N_{kt}, from about 8 to 24, averaging 13. The determination, therefore, is useful only as a preliminary estimate and is not appropriate for design purposes.

For a bearing capacity calculation, Eq. (9.2) can be coupled with piezocone data that are averaged for a depth range from zero to two footing widths below the bottom of a footing. An average effective stress overburden pressure σ_{vo} also can be estimated or determined for that interval. The equation is readily solved with a computer spreadsheet for different depths.

9.10. Precaution with Empirical Relationships

Relying on empirical relationships can be like playing chess with pieces that are subject to mood swings. Cone tests produce three simultaneous measurements that are used to obtain estimates for as many as 10 or more parameters including Poison's ratio, shear modulus, stiffness coefficient, rigidity index, overconsolidation ratio, K_o, friction angle, undrained shear strength, modulus of subgrade reaction, resilient modulus, and California Bearing Ratio. Three test measurements, or degrees of freedom, appear to have inspired an excessive amount of freedom. With the addition of timed data, permeability and a coefficient of consolidation also can be estimated. Relationships still can work if they are truly interrelated. Statistical reliability can be tested by incorporating an unrelated variable to determine if there might be a faux relationship.

9.11. Time-outs for Pore Pressure Dissipation

Penetration by a cone can be stopped to allow excess pore water pressure to drain away, a capability that does not exist with the STP. The rate of decrease can be measured with a piezocone to give a measure of the soil hydraulic conductivity. The data can be reduced by assuming a zone of consolidation of soil close to the cone tip, but there also remains a possibility of influence from radial cracking.

Tests conducted in layered *varved clay* soils revealed that excess pore water pressure diminished 50 percent after 9.5 minutes. Mayne (2007) gives details for the method. As may be expected, pore pressures at the u_1 position on the face of the cone are higher than those at the u_2 position on the shoulder.

9.12. Supplemental Cone Test Data

Various sensors have been added to cone test devices to yield a variety of measurements:

CPT Parameter	Designation	Representative Units
Tip resistance	q_c	Tsf, MPa
Sleeve friction	f_s	Percent
Pore pressure	U_2	Psf, kPa
Shear wave velocity	V_s	Fps, mps
Soil resistivity	R_s	Ohm-cm
Soil temperature	T	°C

Wireless CPT systems eliminate the need for connecting cables, and can speed up the test process. By adding geophones or vibration sensors to the CPT piezocone and providing a sound source, the soil shear wave velocity and multiple soundings can be used to profile a large area and aid in identifying susceptibility to earthquakes. Applications of CPT are discussed by Mayne (2007).

9.13. The Marchetti Dilatometer

A short spade-like instrument with a laterally expandable diaphragm was devised in 1980 by Prof. Silvano Marchetti (1955–2016) of L'Aquila University in Italy. The device (Fig. 9.7) normally is pushed without predrilling, and stops are made at selected depths to obtain pressure readings. Internal gas pressure, usually nitrogen, is used to lift off, release, and re-expand a thin metal diaphragm that is in contact with the soil. The position of the diaphragm is determined by making and breaking electrical contacts. The test is rapid, and the device small enough to enable empirical correlations to be developed using a laboratory pressure chamber. In the field, the Dilatometer usually can be pushed with a conventional truck-mounted drill.

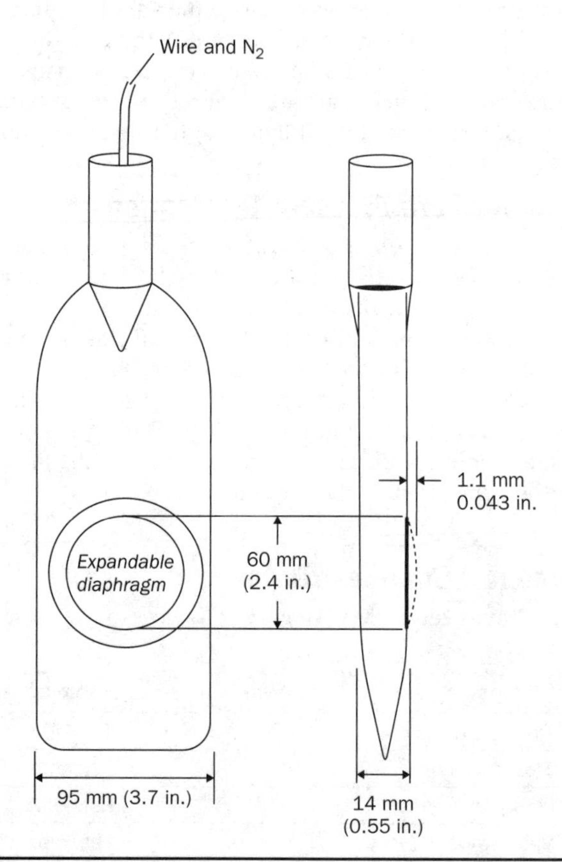

FIGURE 9.7 The Marchetti Dilatometer. The instrument is pushed to a test depth and three different pressure readings are obtained. The instrument then is pushed to the next test depth and the process repeated.

Preparation for Testing

One side of the Dilatometer has an expandable stainless steel diaphragm that is "exercised" by inflating and deflating until the responses stabilize. After the instrument is pushed to a test depth, gas pressure is applied to expand the diaphragm so that it breaks an electrical contact, and the gas pressure is measured. Pressure then is increased until expansion reaches an arbitrary 1.1 mm into the soil and makes a contact, where the pressure again is measured. Gas pressure then is released so that the diaphragm collapses, and then is re-applied to obtain a lift-off pressure indicative of water pressure in the void created by the expansion.

One advantage of the flat Dilatometer geometry is that it relocates tension cracks out to edges (Fig. 9.8). Additional data can be obtained by measuring the force required to push the instrument down into the soil, and additional instrumentation can be added for special purposes, for example, to measure the speed of seismic waves transmitted from the ground surface.

Marchetti defines a *Dilatometer modulus*:

$$K_D = (p_o - u_o)/\sigma_v' \tag{9.3}$$

where p_o is the lift-off pressure, u_o is the static pore water pressure calculated from the depth below a groundwater table, and σ_v' is the calculated vertical effective stress. On

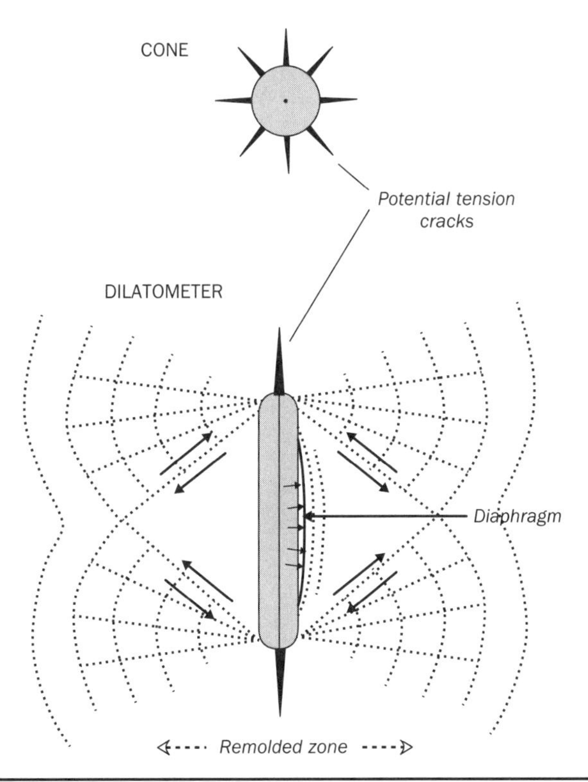

FIGURE 9.8 The flat shape of the Dilatometer moves tensile stress to edges, away from the measurement area of the diaphragm. Stepped Blade tests discussed in the next chapter indicate that the Dilatometer blade thickness sometimes can be sufficient to cause shearing, as suggested in the figure, which probably affects pressure readings.

the basis of laboratory correlations for normally and overconsolidated clays, Marchetti suggests the following relationship for the lateral stress ratio K_o:

$$K_o = (K_D/1.5)^{0.47} - 0.6 \quad (9.4)$$

The relationships have been found to be more applicable for clay soils than sands. More accurate methods for making this determination are discussed in Chap. 11.

Soil Identifications

As no soil samples are retrieved for inspection and identification, Marchetti defines a "material index," I_D:

$$I_D = \frac{\Delta p}{p_1 - u} \quad (9.5)$$

where Δp is the difference between p_0 and p_1, the pressure readings before and after diaphragm expansion, and u is the static pore water pressure. I_D values from 0 to 0.5 are indicative of clay, 0.5 to 1.5 silty soils, and >1.5 sandy soils. Marchetti is careful to point out that this identification is not based on grain sizes but on soil behavior.

9.14. Predicting Settlement

Although foundation bearing capacities should be checked and covered with an adequate factor of safety, the amount or permissible settlement usually is a governing factor for foundation design. A *Dilatometer modulus* is defined on the basis of the pressure required to expand the diaphragm:

$$E_D = 38.2\Delta p \quad (9.6)$$

A vertical *compression modulus*, also called a *constrained deformation modulus*, is:

$$M = R_M E_D \quad (9.7)$$

where R_M is evaluated experimentally from K_D [Eq. (9.3)]. A suggested relationship that is fairly consistent for different soil I_D's is shown in Fig. 9.9.

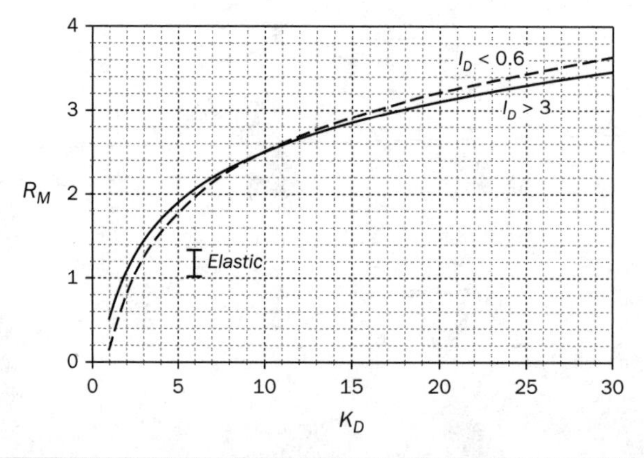

FIGURE 9.9 Determination of R_M for calculation of a vertical modulus, the vertical line indicates a range for ideal elastic behavior. Plotted from equations presented in the *Dilatometer Manual* by Marchetti and Crapps. (Image source: *Geotechnical Engineering: Soil and Foundations Principles and Practice* by Richard L. Handy and Merlin G. Spangler. © 2007, McGraw-Hill Companies.)

9.15. A Key Question: How Can Lateral Yielding Predict Vertical Settlement?

Data presented in Ch. 10 indicate the Dilatometer blade thickness probably is sufficient to cause a bearing capacity failure. That should decrease stress on the Dilatometer. The relatively small movement of the diaphragm is unlikely to re-initiate shearing so it can be speculated that it may give a measure of compression of soil in the remolded zone. The strength reduction may be slowly recoverable through *aging*.

A statistical verification of the use of Dilatometer data to predict settlement is shown in Fig. 9.10, as it gives a respectable R^2 value. However, a much higher value is required for design, indicated by the lines enclosing data in the figure. (R^2 for Borehole Shear Test data usually exceeds 0.99.)

Aging

Aging that occurs in geological time can transform sand into sandstone, and slow reorientation of clay particles can turn clay into shale. Aging can begin with minor rearrangements in contact bonding that removes weak spots. The effect of aging on strength is indicated by Dilatometer modulus data shown in Fig. 9.11. Both sets of data may be affected, but the larger reduction in modulus occurred in the fine silty sand soil after it had been excavated for a trench and re-compacted for backfill. A strength recovery through aging should tend to put Dilatometer interpretations on the safe side for design.

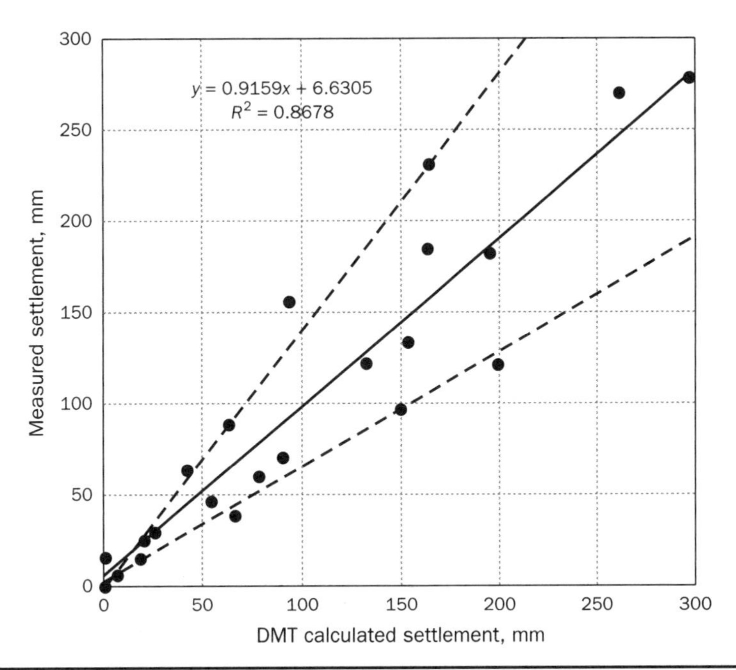

Figure 9.10 Statistical verification that Dilatometer data can predict settlement with a range in error of about +/−30 percent. (Redrawn from Marchetti, 1997, using data from Hayes. Schmertmann (1986) reports a range from −29 to +67 percent.

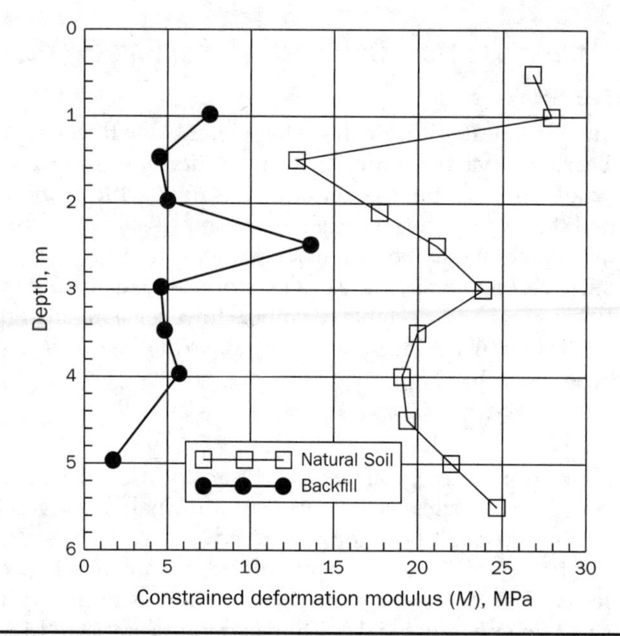

FIGURE 9.11 Dilatometer data showing a loss of strength that can be attributed to the soil structure and which presumably may be recovered through aging. The soil is silty sand. (Data courtesy of Roger Failmezger.)

A Dilatometer Shift in Direction of the Major Principal Stress

In normally consolidated soil the major principal stress direction is vertical, whereas in soil close to the Dilatometer it most likely is horizontal. Diaphragm expansion therefore works against the localized major principle stress direction, simulating vertical compression of soil underneath a foundation. As the response to the diaphragm is likely to have been influenced by remolding it may tend to result in overdesign. Generally the larger and more costly the project, the greater the justification for more accurate data acquisition and analysis.

The reduction of Dilatometer data is described in more detail in instructions supplied with the instrument.

Problems

1. Estimate the angle of internal friction for soil at a depth of 10 m in Fig. 9.5. How reliable is the estimate?
2. What variables may influence piezocone pore water pressure determinations?
3. Is the "friction ratio" as commonly used a measure of friction? Explain.
4. Estimate cohesion for the clay in Fig. 9.5. How reliable is the estimate?
5. Why should seismic shear wave velocities be more diagnostic of poor soil conditions than compression wave velocities?
6. How may a strength gain through aging affect designs based on Dilatometer data. May the same process affect the use of cone and SPT data for design?

References

Marchetti, Silvano., *The Flat Dilatometer: Design Applications*. Proc. of the Third Getoechnical Engng. Conf., Cairo University. 423–448, 1997. *(Available online)*.

Mayne, P. W., *Cone Penetration Testing*, NCHRP Synthesis 368, Transportation Research Board, Washington, DC, 2007. (Available online.)

Schmertmann, J. H., Hartmann, J. P., and Brown, P. R., *"Improved Strain Influence Factor Diagrams," ASCE J. Geotech. Engr. Div.*, 104:1011–1043, 1978.

Schmertmann, J. H., *Dilatometer to Compute Foundation Settlement*. Use of In Situ Tests in Geotechnical Engineering, ASCE, 303–321, 1986.

Further Reading

Failmezger, R. A., and Anderson, J. B., eds., *Flat Dilatometer Testing*. Proceedings of the Second International Conference on the Flat Dilatometer, Washington, DC, 2006. (Available online.)

CHAPTER 10

Focus on Lateral Stress

Lateral Confining Stress can Keep Soil from Squeezing Out from Underneath a Foundation

10.1. Lower Cost, More Convenient

Although foundations apply vertical stress, tests described in this chapter apply stress horizontally, in part because a reaction base is readily available on the opposite sides of a borehole. Two of the tests described in this chapter translate horizontal into vertical responses through the use of empirical correlations. A third test focuses on rapid and accurate measurement of horizontal in-situ soil stress because of its major role in foundation engineering.

10.2. The Pressuremeter

In 1933, German engineer F. Köhler described a test device with an inflatable rubber bladder that was expanded inside a borehole to obtain a horizontal stress–strain relationship. However, inflation was with gas pressure, so pressure–volume relationships were sensitive to temperature. Further investigations probably were interrupted by World War II.

In the 1950s, French engineer Louis Menard used gas pressure over water to inflate the bladder, and obtained a more accurate measurement of volume changes by measuring the water level in a calibrated chamber. Another significant improvement was the addition of separate inflatable "guard cells" at both ends of the bladder, in order to ensure that soil expansion is directed horizontally. This is depicted in Fig. 10.1. In 1956, Menard wrote his thesis on the device at the University of Illinois, perhaps one of the more insightful MS theses ever written.

Soil Disturbance from Drilling

Relieving lateral restraint on soil with an open boring can allow expansion and change the soil structure, so French practice is to bore holes with drilling mud to maintain fluid pressure on the soil. The mud is circulated to bring up drill cuttings for soil identifications. Cuttings are allowed to settle out in a "mud pit," so the mud can be reused.

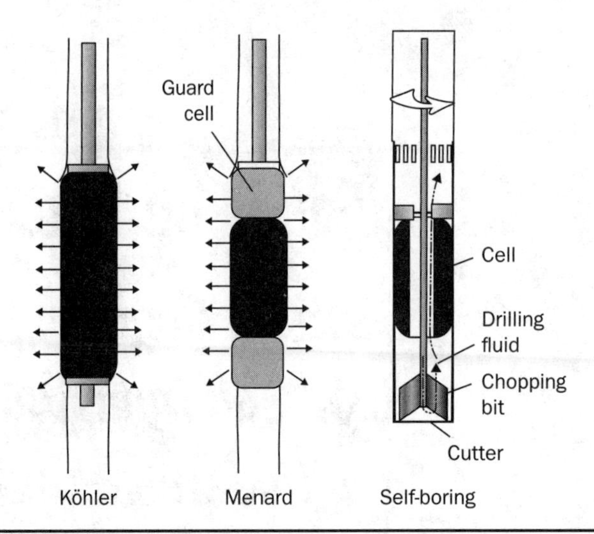

Figure 10.1 Steps in development of the pressuremeter. (Image source: *Geotechnical Engineering: Soil and Foundations Principles and Practice* by Richard L. Handy and Merlin G. Spangler. © 2007, McGraw-Hill Companies.)

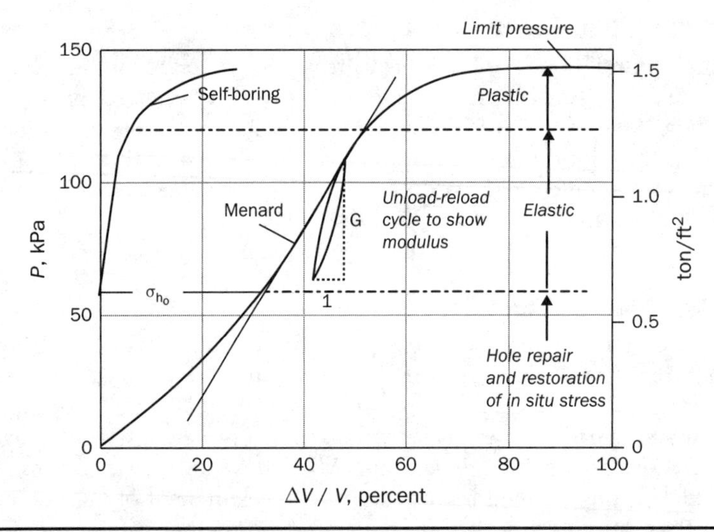

Figure 10.2 Representative Menard Pressuremeter and self-boring pressuremeter data. (Graph source: *Geotechnical Engineering: Soil and Foundations Principles and Practice* by Richard L. Handy and Merlin G. Spangler. © 2007, McGraw-Hill Companies.)

Self-Boring Pressuremeters

Self-boring pressuremeters were developed in France and England, and also used drilling mud. As depicted in Fig. 10.1, a cutting bit is surrounded by a steel tube that is advanced as the soil is being cut away, the objective being to slide the tube down into a boring with minimum disturbance to the soil.

Comparative data in Fig. 10.2 indicate that this objective can be accomplished in fine-grained soils. Coarse particles must be either plucked out or pushed aside into the soil (Ladd, 1980). Tests also are considerably more time-consuming than those made with a conventional pressuremeter and power drilling.

10.3. Interpretation of Pressuremeter Test Data

Lateral In Situ Stress

Data in Fig. 10.2 indicate that expansion of a self-boring pressuremeter begins when the expansion pressure exceeds the lateral stress existing in the soil, designated as σ_{ho}. Lateral stress is relevant to friction on pile foundations, and a value that exceeds the vertical stress can provide field evidence for overconsolidation. Lateral in situ stress also can be estimated with a conventional pressuremeter, by cycling the expansion pressure as shown in the figure. An important limitation is because bladder expansion will seek out the direction of lowest lateral stress, not the highest that is more relevant for bearing-capacity failures and friction on piles.

The Limit Pressure in Foundation Engineering

Figure 10.2 shows a maximum lateral *limit pressure* that can be sustained by the soil and is the same for both types of pressuremeter. As the limit pressure tends to stay constant, it may reflect the initiation and growth of radial tension cracking shown in Fig. 10.3 II. Soil between open cracks may fail in unconfined compression to create a remolded zone, Fig. 10.3 III. This possibility is supported by lateral stress measurements in soil at various distances from Rammed Aggregate Piers®, at stress levels that are comparable to those created with a pressuremeter. These data are discussed in Chap. 12.

When soil between radial tension cracks fails in unconfined compression, the failure zone should relocate outward where stress is less concentrated. Expansion may continue unless it is stopped by reducing the pressure. It usually is stopped when $\Delta V/V$ reaches 100 percent in order to prevent separation of thin overlapping metal strips that often are used to protect the rubber bladder of a pressuremeter.

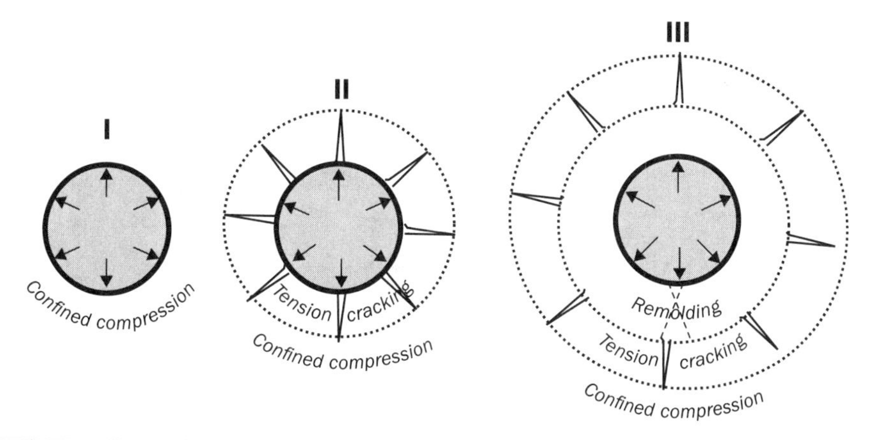

FIGURE 10.3 Radial compression and hypothesized possible tangential tension cracking of soil near an expanding pressuremeter.

Radial stress applied by a pressuremeter must first exceed the existing K_o lateral stress in order to create tension, and linear relationships between limit pressure and depth are revealed in data reported by Baguelin et al. (1978). That trend also is consistent with the depth factor in the Terzaghi bearing capacity equation.

A Theoretical Approach

With increasing radial distance outward into the soil, radial compressive stress decreases and tangential tensile stress increases, as shown in Fig. 10.4. The lateral unconfined compressive strength qu therefore should defined a sheared and remolded soil zone close to the pressuremeter. Similarly the distance where tangential stress becomes positive should define the extent of tension cracking.

In 1852, French mathematician Gabiel Lamé (1795–1870) presented an equation for radial expansion of a vertical cylinder in an elastic medium, but without the influence from possible tension cracking:

$$G = V\frac{\Delta p}{\Delta V} \tag{10.1}$$

where G is the shear modulus. Menard suggested that V can be represented by the average volume at ends of the linear relationship and designated the ratio as G_m.

From theory of elasticity, G relates to the elastic modulus E as

$$E_p = 2G_m(1 + \upsilon) \tag{10.2}$$

where E_p is the *pressuremeter modulus* and υ is Poisson's ratio. Menard combined the equations:

$$E_p = 2(1 + \upsilon)\,(V_c + V_m)\,\frac{dp}{dV} \tag{10.3}$$

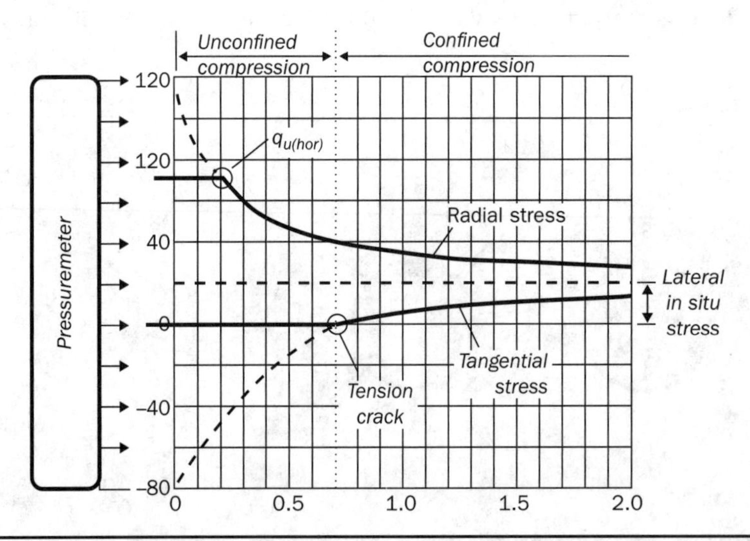

FIGURE 10.4 Elastic distribution of pressuremeter-induced radial (upper curve) and tangential (lower curve) stresses in a horizontal plane.

Clay			Sand		
psi		kPa	psi		kPa
0–10	Very soft	0–75	0–30	Very loose	0–200
10–20	Soft	75–150	30–75	Loose	200–500
20–50	Firm	150–350	75–220	Compact	500–1,500
50–110	Stiff	350–800	220–360	Dense	1,500–2,500
>230	Hard	.1,600	>360	Very dense	>2,500

Source: Baguelin et al. (1978).

Table 10.1 Soil Properties Estimated from Pressuremeter Limit Pressure

where υ is Poisson's ratio, V_c and V_m are pressuremeter volumes at ends of the linear relationship, and dp/dV is the change in pressure per unit change in volume.

As the above theoretical development does not incorporate possible tension cracking, and Menard indicates that Eq. (10.3) overpredicts foundation settlement. Baguelin et al. (1978) compare pressuremeter and conventional methods for settlement analysis, and conclude that instead of taking a one-dimensional, volume-change modulus for the soil (i.e. the Terzaghi method) and adjusting it for lateral yield of the soil, the pressuremeter method starts with a distortion (lateral yield) modulus and adjusts it for volume change (consolidation).

Consolidometer data do not include lateral yielding of the soil and therefore may tend to underpredict settlement, especially for narrow foundations.

Use in Design

Settlement predictions developed empirically from pressuremeter data incorporate footing sizes and depths and can become quite complicated, and it is suggested that design procedures be followed that are recommended by equipment manufacturers. In all cases follow-up measurements should be made to test accuracy.

Soil Identifications

Soil types can be directly identified from soil borings, but pressuremeter data can be more indicative of engineering properties of the undisturbed soil. Some observations are summarized in Table 10.1.

10.4. The K_o Stepped Blade

Lateral soil stresses that exist or are developed in situ directly affect friction on piles. Lateral stress also is relevant to shallow foundation bearing capacity, because it is not only the soil underneath a foundation that holds the building up; it also is restraint from soil alongside that keeps the soil underneath from squeezing out.

Engineers at the U.S. Department of Transportation, Federal Highway Administration (DOT-FHWA) recognized a need for a rapid, accurate test to measure

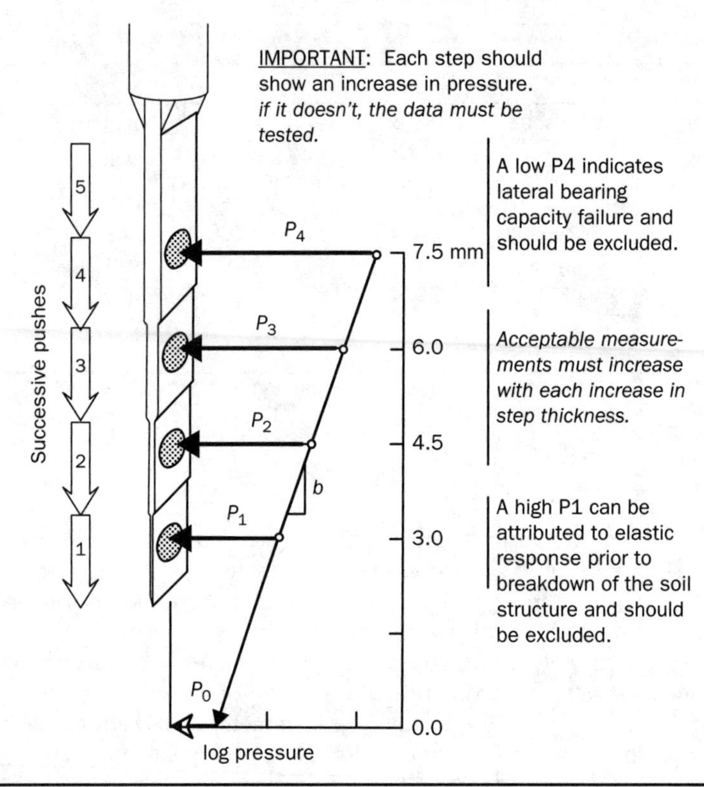

FIGURE 10.5 Concept of soil lateral pressure measurement with the K_o Stepped Blade. The blade is advanced between readings, and data sets are assembled for each subdepth. (Illustration and graph source: Courtesy of Handy Geotechnical Instruments, Inc.)

lateral in situ stress. Since accuracy of pressuremeter tests depends on boring a hole without disturbing soil around the hole, another approach is to intentionally introduce different levels of disturbance and extrapolate data to obtain a hypothetical value with zero disturbance. A diagram of the K_o *Stepped Blade* is shown in Fig. 10.5. The blade has four arbitrary step thicknesses, 3, 4.5, 6, and 7.5 mm. A hollow rib on the back adds stiffness and contains pressure lines used to activate the pressure cells.[1]

The Two-Chambered Pressure Cell

The thin blade steps required the development of a new kind of pressure cell with pneumatic operation so that pressures can be directly measured without the need for

[1]Many students were involved in the development and use of the Stepped Blade. The final design was by Don Eichner. The extended thickest step adds an additional data set and was suggested by Scott Mackiewicz.

calibrations. Each cell has two chambers that are separated by a flat ridge that is covered with a thin diaphragm that is in contact with the soil. Gas pressure is increased to both parts of the cell but with a small differential pressure, which drops to zero when the cell pressure equals the soil pressure. The gain in gas pressure temporarily is halted, pressure is read from a pressure gauge, and pressures immediately are vented to prevent damage to the diaphragm. A reading requires about one minute, after which valves are switched to read the next cell. Teflon® diaphragms proved to be most durable. Air or nitrogen usually is used to expand the diaphragm.

Test Sequence

Push one step length and read Cell 1; then push another step length and read 2-1; then another step length and 3-2-1, etc. The order maintains essentially the same drainage time for each subdepth, and excess pore water pressure also extrapolates out. The procedure is repeated at five subdepths, which requires about 15 minutes. The instrument then is removed and the boring extended to the next test depth. *Unlike the cone and Dilatometer, it must not be pushed continuously.*

Interpretation

The logarithm of pressure, log *P,* is graphed against cell thickness (Fig. 10.6), giving a graph that is analogous to a linear e-log *p* determination from consolidation tests.

Three soil behaviors are typical:

1. A *high* first data point is attributed to pressure required to break down the soil structure, and *is not included* in the analysis of lateral in situ stress.
2. A straight-line relationship on a semilogarithmic graph is used to determine lateral in situ stress. As depicted in Fig. 10.6, data from this relationship can be extrapolated to obtain a hypothetical pressure on a zero-thickness blade.
3. A *low* final data point can indicate a lateral bearing-capacity failure and *is not included* in the analysis. (This observation also is relevant to the interpretation of both cone and dilatometer test data.)

The total number of pressure measurements in a single test series is shown in Table 10.2. Each successive data point is tested, as it must show an increase in pressure. If it does not, one or the other data point must be rejected. Only two data points are required to establish a relationship if the slopes are consistent. First, second, and third subdepths are most likely to yield interpretable data. As shown in Fig. 10.6, the duplication of results at nearly the same depth can add credence to the measurements, and results that are in close agreement may be pooled and treated statistically to obtain means and standard deviations.

Example

Representative K_0 *Stepped Blade* test data for the first three subdepths are shown in Fig. 10.6. The single high data point occurred with the thinnest step, prior to breakdown of the soil structure. As the test was conducted above the groundwater table, the graph slope should give a measure of soil stiffness that increases with depth because of changes in the soil *B horizon.* The extrapolated lateral stress remains essentially the same.

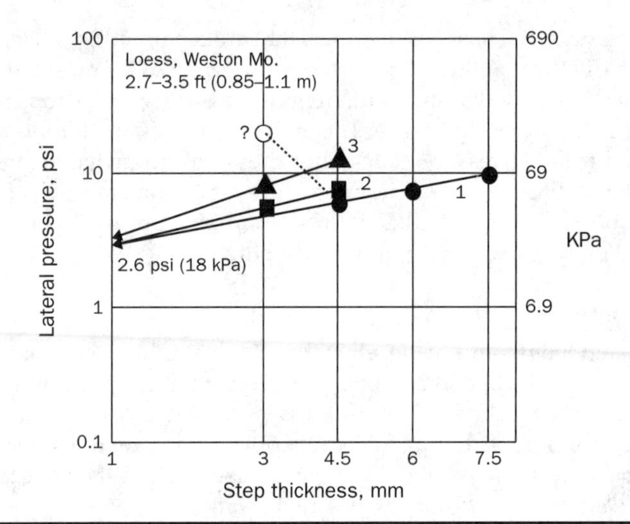

FIGURE 10.6 K_0 Stepped Blade data from subdepths 1, 2, and 3 from a moderately expansive weathered soil profile. One data point is high, prior to breakdown of the soil structure, and is not included in the extrapolations. Close agreement between the three determinations is indicative of accuracy. (Graph source: *Geotechnical Engineering: Soil and Foundations Principles and Practice* by Richard L. Handy and Merlin G. Spangler. © 2007, McGraw-Hill Companies.)

	Pressure Measurements			
Subdepth	**3 mm**	**4.5 mm**	**6 mm**	**7.5 mm**
First	•	•	•	•
Second	•	•	•	•
Third	•	•	•	
Fourth	•	•	(Limited interpretation)	
Fifth	•		(No interpretation)	

TABLE 10.2 Maximum number of SB Data Points

The K_0 calculated from the depth and an assumed soil unit weight is about 1.3, indicating expansive clay that is consistent with the observed soil structure.

10.5. Summary

This chapter describes three lateral in situ soil tests: two intended for site evaluations, and the third is designed to measure lateral in situ soil stress and therefore can provide insights into soil behaviors involved in the other two. The third test is relevant for understanding processes involved in bearing capacities of deep foundations and pile group action (discussed in Chapter 11), and for defining stress-related methods for soil improvement (Handy, 2013, and Chapter 12).

Problems

10.1. In Fig. 10.2 why is an unload-reload cycle used to measure elastic response with the Menard Pressuremeter and not with the self-boring Pressuremeter?

10.2. (a) How can one estimate lateral in-situ soil stress with the Menard Pressuremeter?

(b) If the pressure is different in the x and y directions (z being up), which is more likely to be measured?

10.3. What is Poisson's Ratio, and why is it required to predict foundation settlement from Pressuremeter data?

10.4. Poisson's Ratio ordinarily is in a range 0 (for low-density) to 0.5 (for high-density) soil, and even higher for dilating soils.

To what extent can Poisson's Ratio be expected to affect the Pressuremeter modulus, Ep?

10.5. How can one measure Poisson's Ratio?

10.6. Why is an empirical correlation required to predict foundation settlement from Dilatometer data?

Why is it not required when using consolidation test data?

10.7. What are two approaches to direct measurement of lateral stress in soils?

(a) Suggest some areas where the information can be useful in foundation engineering.

(b) Are there any advantages to one or the other of the approaches? Which has been in use longer?

References

Baguelin, F., Jezeguel, J. F., and Shields, D. H., *The Pressuremeter and Foundation Engineering*, TransTech Publications, Clausthal-Zellerfeld, Germany, 1978.

Handy, R. L., "Liquefaction, Lateral Stress, Consolidation State, and Aging," *ASCE Journal of Geotechnical and Geoenvironmental Engineering*, 139(10), 1808–1812, 2013.

Ladd, C. C., *Evaluation of Self-Boring Pressuremeter Tests in Boston Blue Clay.* Report No. FHWA-RD-80/052 of the Federal Highway Administration, Washington, DC, 1980. (Available online.)

Graduate student inserting the Stepped Blade into a test boring next to a drilled pier, Savanah, Ga. Soil pressures are measured on four different step thicknesses and graphed to obtain a pressure with zero thickness. After measurements at different sub-depths, the Blade must be removed and the boring advanced for the next test series. Author photo.

CHAPTER 11

Design of Deep Foundations

Out of Sight but Not Out of Mind

11.1. Transferring a Foundation Load Deep to Reduce Settlement

Figure 11.1 shows results from a consolidation test on a clay soil, where the sloping line shows a change in void ratio that is linear to the logarithm of the applied pressure. In this pressure range the soil is said to be "normally consolidated."

The logarithmic scale means that a foundation load that is applied at a shallow depth, as at point A, should cause a significantly larger change in the soil void ratio and settlement than if the same load is transferred downward to a depth represented by point B. In this case transferring the foundation load twice as deep can reduce the change in void ratio from about 0.1 to about 0.05, a reduction of about one-half. Settlement therefore can be significantly reduced even when loads are supported by normally consolidated soil where the soil density is in equilibrium with the overburden pressure.

An alternative objective that often is used to support heavier loads is to transfer them all the way down to bedrock and minimize future settlement.

11.2. When Pile Foundations Became a Matter of Necessity

In about 4,000 BCE, ever-popular lakeside properties in Switzerland suffered the inconvenience of seasonal flooding from glacial melting so residents started supporting their houses on wood posts. Ironically, the same procedure is being used today, in response to a rising sea level and bigger storm waves. Posts that are driven down into the ground are *pile foundations.* They engage soil like friction engages a nail.

11.3. Soils and City Planning

Tall buildings occupy less space and obviously are heavier, but when options are to walk long distances or ride in an elevator, most people prefer the elevator. Thanks to the

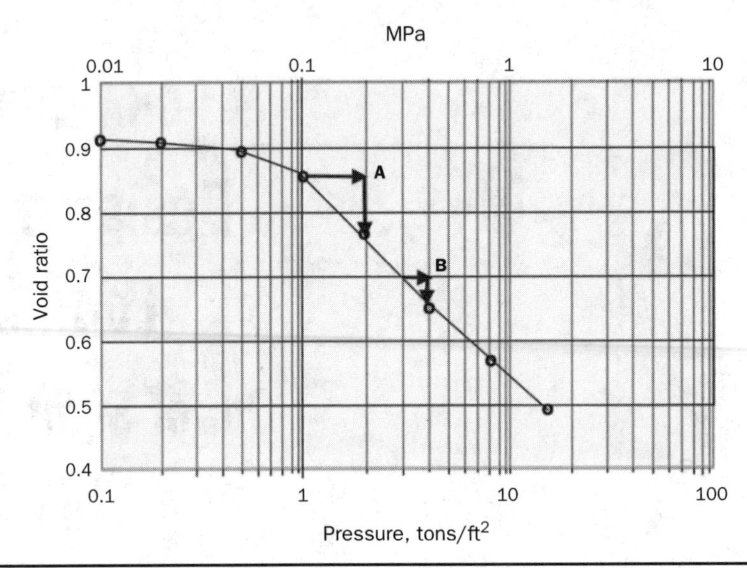

FIGURE 11.1 The logarithmic pressure scale means that moving 1 Tsf added stress deeper, from A to B, in a normally consolidated soil can reduce the change in the soil void ratio and reduce settlement.

inventiveness of Elisha Otis, elevators have automatic braking, and it also is unlikely that anybody ever will be killed in a head-on collision between two elevators.

Cities and Rivers

Cities often were founded where there was access to transportation by water, so there usually is a river running through the middle. Cities therefore require bridges that are supported on alluvial, river-deposited soils. Many buildings are likely to be supported on the same kinds of soils. Bridges and heavy buildings are likely to be supported on deep foundations.

Cities can be laid out like a wagon wheel laid flat, with a high-rise hub in the center and spokes extending outward to connect to suburbs. Cities located by the sea are more likely to form a half-circle with spokes extending outward from a fish market.

11.4. Lowering of Sea Level

As discussed at the beginning of this book, continental glaciers captured and held so much of the Earth's supply of water that sea level was lowered about 400 ft (125 m). That caused major rivers worldwide to entrench, to cut deeper and in the process also widen their valleys. When glaciers melted, sea level started to rise, and sand and gravel carried by glacial meltwaters filled the valleys. They then were covered under an assortment of alluvial soils but still play a prominent role for support of deep foundations. The sand and gravel strata are so thick that it usually is not feasible to extend foundations all the way down to bear on bedrock.

11.5. End Bearing

There are so many unknowns that the use of soil mechanics to predict deep foundation bearing capacities can be like trying to catch a butterfly with a fishing pole. That does not mean one should not take a swing at it. Terzaghi's shallow foundation bearing capacity equations usually is a starting point, even though the failure geometries are quite different. As shown in Figs. 5.8 and 7.3, slip surfaces for shallow foundations extend up to the ground surface. That almost certainly does not occur with deep foundations. Instead the geometry is more likely to resemble that created by pushing a cone (Fig. 9.4). In addition the classical derivations are two-dimensional, and punching failure of a deep foundation is three-dimensional.

The "N" factors in the Terzaghi bearing capacity equation are: N_γ, which is multiplied by a foundation width and is designated as a width factor; N_c, which is multiplied by the soil cohesion and is a cohesion factor; and N_q, which is multiplied by depth and is a depth factor. Most important for deep foundations is the surcharge factor, N_q, to the extent that the other factors are conveniently ignored.

Various solutions for N_q have been summarized by Prof. Aleksandar Vesic (1924–1982) at Duke University and reported by many investigators including Lambe and Whitman (1969). They show that with a soil angle of internal friction of 35 degrees, the value for N_q may vary from 3.2 to 630, a 200-fold difference. It therefore should come as no surprise that pile load tests are frequently used to validate designs.

Most engineers prefer to underpredict a bearing capacity so it passes a load test, instead of overpredict it; therefore, they are required to make a difficult decision, to simply improve the design or improve the design and retest, increasing cost and delaying construction. One also can try and justify accepting a marginally lower factor of safety, but always with the owner's approval. A case history is cited in the text following Fig. 11.3. If a test does not support the design load, that obviously restricts the options.

11.6. Pile Driving

Wood Piles

At one time straight trees were reserved for ship masts. That use has declined, but they still are used for piles. Trees are stripped of bark and branches, and may be chemically treated to protect from decay. Decay also is prevented by permanent submergence under oxygen-deprived water.

Wood piles normally are driven upside down to take advantage of a natural taper and prevent making an oversize hole. Taper also is employed in some manufactured steel or concrete piles, in order to achieve a more uniform distribution of side resistance along the length of a pile.

The Science of Hammering

Pile driving can be conducted by repeatedly lifting and dropping a guided hammer with a crane. The hammer also can be lifted and dropped using steam or air pressure. Diesel hammers perform like a piston in a diesel engine: They draw in air and fuel on the upstroke, and then compress the mixture on the downstroke so it is heated sufficiently that it explodes. The explosion drives the bottom of the cylinder down and simultaneously raises the hammer back up, so it draws in more air and fuel and starts the next cycle. Diesel hammering can continue until the fuel is shut off, or it can be interrupted and must be restarted if driving encounters a soft zone or cavity.

Hard Driving and Brooming of Wood Piles

Impacts from driving can be damaging to both ends of a wood pile. Lower ends can be protected with pointed steel attachments, but hammering on the top can cause splitting or "brooming." Temporary "cushions" of green wood or fir plywood are commonly used to take the impacts. Cushions can increase the time required for driving, and the cushion must be replaced if the energy conversion causes it to catch on fire or be reduced to charcoal.

No Lunch Breaks!

If pile driving is stopped for more than a few minutes, a pile can "set" so it is difficult or impossible to restart. A pile that does not go to a full design is rejected, but as pulling can disturb the soil, it may be cut off and bracketed by a pile on both sides, and linked with a reinforced concrete cap.

11.7. Tension Breaks in Concrete Piles Caused by Pile Driving?

After some driven concrete piles failed pile load tests, the piles were pulled and found to have horizontal tension cracks. Researchers at Texas A & M University suspended test piles horizontally on cables and hammered on one end, and detected a compression wave that traveled the length of a pile until the tip extended and then rebounded to create an echo wave that went back through the pile. At the point where the return wave passed the compression wave from the next hammer blow, the pile broke in tension. This can be prevented with longer cycle times for hammering, but that may not be possible with a diesel hammer. Thicker wood cushioning can slow the impact times. Tension also can be reduced by steel reinforcing. A tension break can be detected in an installed pile with sound waves that created a premature echo.

Piles Doing a U-turn

On rare occasions a driven steel pile will strike a glancing blow against a boulder and be bent so it does a U-turn and comes up out of the ground at a distance away from the pile driver to the amusement of onlookers. Even though that is unusual, it can be a hint that other driven piles may bent and the bend not detected. Such a pile is likely to fail a load test, but not all piles are load-tested. This possibility favors the use of piles in groups connected at the top with a cap of reinforced concrete.

11.8. The Engineering News Formula

In 1888 a formula relating pile bearing capacity to impact energy was presented by A. M. Wellington in *Engineering News* (later the *Engineering News Record*). He calculated the hammer energy from weight times the drop distance and equated it to the pile incremental movement, then added a factor of safety of 6 to account for energy losses.

The formula presents a rational approach, and despite the arbitrary nature of the factor of safety it became widely used:

$$Q_p = \frac{1}{6}\frac{WH}{S} \tag{11.1}$$

where Q_p is the allowable bearing capacity (tons), W is the hammer weight (tons), H is the drop height (feet), and S is the penetration distance per blow (feet).

11.9. Pile Bearing Capacities and Load Tests

The variability of load capacities indicated in Table 11.1 can justify the use of full-scale load tests. One or more production piles or piers may be subjected to a load test, where the maximum test load is usually two times the anticipated final load. If a test passes, that indicates a *factor of safety* of 2 or more. The factor of safety can only apply to the pile or pier being tested, but also can imply an acceptable *margin of safety* for an entire foundation system.

Strength Gains and Slow Loading

A waiting period of a week to a month after installation is required for concrete to set and harden, and to allow soil disturbed by boring or pile driving to recover strength. Loading then is slow and incremental to encourage drainage of pore water and reduce the amount of excess pore water pressure. Creature comforts usually include shade from a canopy that is required to prevent heat from the sun from causing thermal distortions that can affect measurements. Measurement devices, dial gauges and transducers, are independently supported on cross-beams to avoid interactions with soil close to the test pier or pile.

Anchor Requirements

A setup for a static load test of a large-diameter concrete test pier is shown in Fig. 11.2, where the cross-beam held down by being attached or welded to reinforcing steel extending out of the tops of anchor piers at each end. The anchors are in tension and are held down by skin friction and cohesion along the sides, so the sum of the side restraint from two anchors must exceed the sum of skin friction plus end bearing of the test pile or pier.

Type	Material	Typical Capacity, tons (kN)
Timber	Treated wood	10–50 (100–500)
Steel	Heavy wall pipe	30–150 (300–1,500)
	H-section	40–120 (400–1,200)
	Fluted shell	40–150 (400–1,500)
Concrete	Drilled pier (caisson)	50–70* (500–700)
	Precast	40–100* (400–1,000)

* Limited to 33 percent of the 28-day strength of the concrete.

Table 11.1 Some Representative Ranges in Load Capacities for Deep Foundations

FIGURE 11.2 Load test setup for a concrete pier at the center, with the cross-beam held down by steel-reinforced tension piers at both ends. (Image source: *Geotechnical Engineering: Soil and Foundations Principles and Practice* by Richard L. Handy and Merlin G. Spangler. © 2007, McGraw-Hill Companies.)

Conduct of a Test

Downward force on the test foundation is applied incrementally with a hydraulic jack, with each increment held constant to allow time for the system to reach equilibrium. On-site graphing of data can reveal progress of a test.

Criteria for Failure

A "plunge" is shown at the right in Fig. 11.3, and can be divided by a factor of safety for design. A common design factor of safety based on plunge is 2, which in this test would give a maximum working load of 95/2 = 47.5 tons.

An alternative requirement may be specified, for example, that settlement must not exceed an arbitrary amount that typically is 1 in. (25 mm). That requirement is shown at point *H* in the figure, and is not applicable because the load is higher than that defined by the plunging failure.

Another criterion that may be useful for factories and warehouses with highly variable loading can be based on an allowable amount of deflection per unit load. For example, in Fig. 11.3 a slope of 0.05 in./ton can define *OA*″ and a failure at point *F*. In that case the most critical requirement still is the amount of load that caused the plunging failure.

Marginal Designs

When a load test does not generate a required factor of safety, some common options that may be considered are as follows: (a) increase the length and conduct another load test, (b) increase the length without conducting another load test, and (c) accept the lower factor of safety. An owner may argue for option (c) in order to save time and money up-front, but failure under a structure is not an option. An important criterion is the soil variability that can be revealed by exploration drilling.

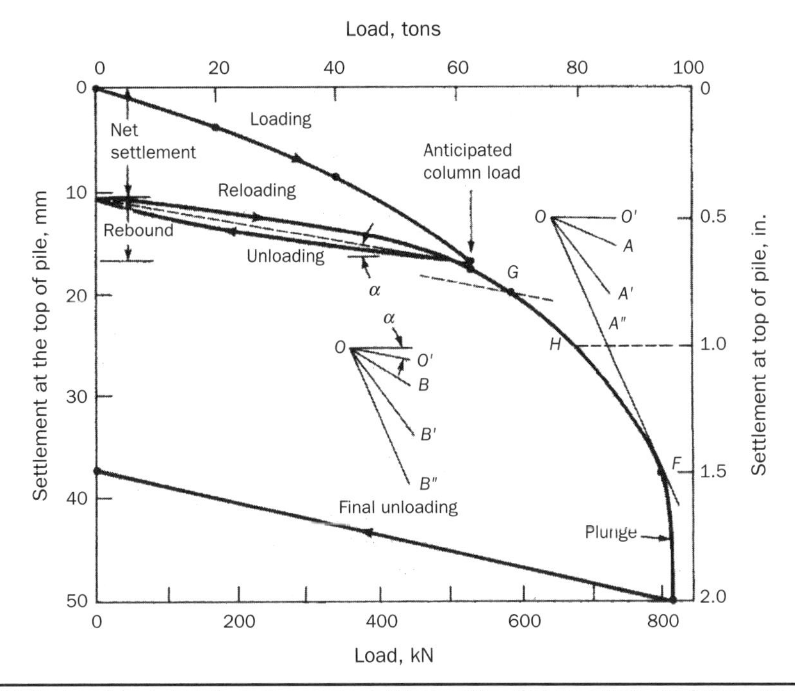

FIGURE 11.3 Load test data from the previous figure, with arbitrary criteria for failure. (Image source: *Geotechnical Engineering: Soil and Foundations Principles and Practice* by Richard L. Handy and Merlin G. Spangler. © 2007, McGraw-Hill Companies.)

Case history. The pier in Fig. 11.3 was installed in glacial till that had been thoroughly mixed and overconsolidated under the weight of glacial ice. Time was an issue and the margin was small, and the owner opted for option (c) without retesting. The building was completed on time and there were no settlement issues.

11.10. Analyzing Hammer Blows

A Wave Equation for Driven Piles

In 1960, E. A. L. Smith of the Raymond Pile Co. proposed a model consisting of pile segments separated by springs (Fig. 11.4). Elastic constants are applied to define the amount of compression of each segment as a function of time, and the model can replicate rebound and opposing compression waves that can contribute to tension cracking. *However, a model that is based on the dynamics of driving does not incorporate setup factors that can contribute to a pile bearing capacity, particularly in clay soils.*

A Pile Driving Analyzer (PDA)

Newton's Second Law, $f = ma$, can be applied to pile driving by attaching an accelerometer near the top of a hammer-driven pile. Acceleration a in the pile is negative as

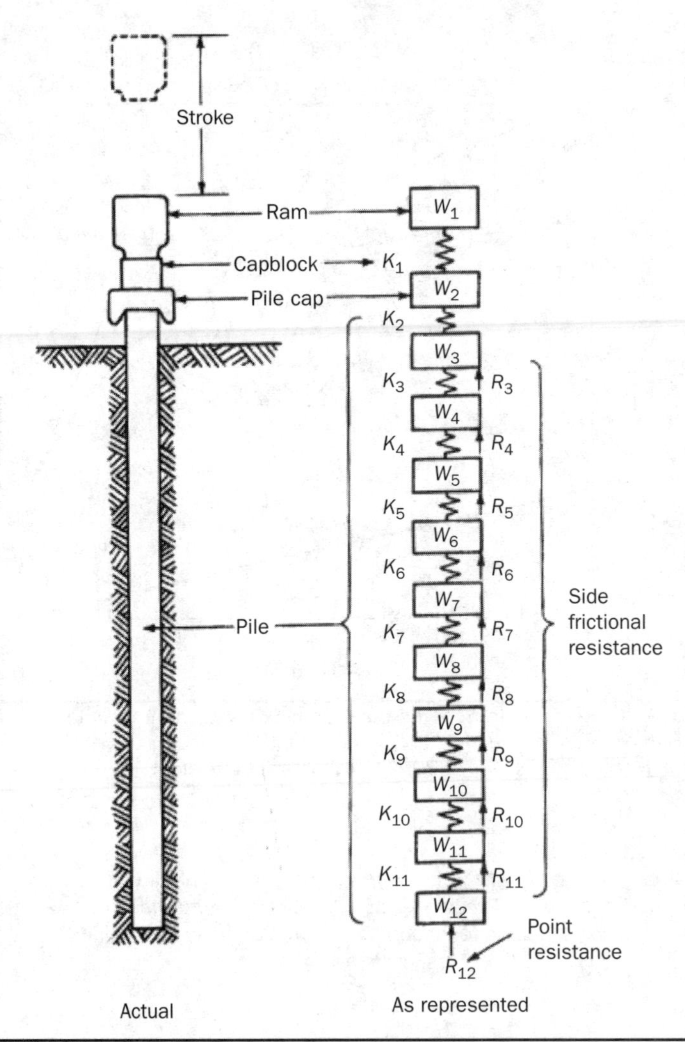

Figure 11.4 Wave equation model for driven piles. (Image source: *Geotechnical Engineering: Soil and Foundations Principles and Practice* by Richard L. Handy and Merlin G. Spangler. © 2007, McGraw-Hill Companies.)

it decreases following each hammer impact. Energy relationships can be influenced by damping, which tends to be higher for clay.

Measuring Setup with PDA and Restrike

Research conducted at Case Western Reserve University adapted the PDA restrike method for measuring the response of installed piles and drilled shafts. The process also is called a High Strain Dynamic Test, and is described in ASTM Designation D4945. *It should be noted that in order to create a useful response, the restrike hammer blow must move the pile.* A plywood cushion still is recommended during driving.

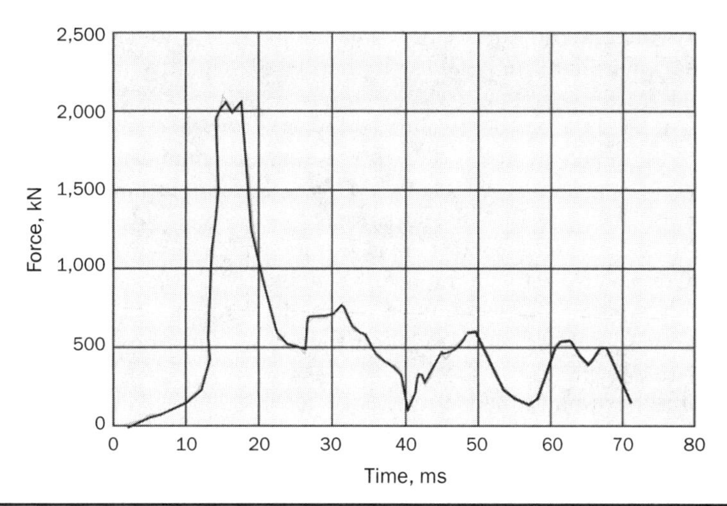

Figure 11.5 Wave arrivals detected with PDA, a pile driving analyzer. (Graph source: Adapted from FHWA, Project Publication No. FHWA-HRT-05-159, 2006.)

An example of a PDA is presented in Fig. 11.5. The analysis was made using the Case Pile Wave Analysis Program or CAPWAP®. The method gives an estimate for the total bearing capacity, and also can reveal a discontinuity by detecting an early echo. Interpretation of PDA data and results from the CAPWAP usually are performed by a service group specializing in their use.

11.11. Citizen Complaints

Continuous, day-in day-out hammering can cause discomfort, and vibrations can be amplified if the end of a driven pile encounters a hard object such as buried slab of concrete. If horizontal brick lines in a nearby building start puffing dust, citizen complaints and a potential lawsuit may become inevitable. Before-and-after photographs can aid in assessing damage, and often can be agreed to by owners of adjacent structures.

Vibratory pile drivers avoid impacts and are less noisy. The vibrator usually consists of two side-by-side counter-rotating wheels with off-center weights. Vibrating pile drivers are most effective with granular soils that have a moderate to high void ratio, but also can be used in finer grained soils and clays if they are not highly overconsolidated.

11.12. Pile Load Capacities: End Bearing

End Bearing on Rock

Given a choice between deep foundations bearing on soil or on solid rock, most engineers prefer rock, but not all rock is solid.

The most abundant sedimentary rock is shale, which is thin-bedded, has a high clay content, and is relatively weak and easily penetrated by driven piles. Claystone and mudstone are similar in composition, but without the thin bedding so they are more resistant. Sandstone and limestone are common and much more resistant, but it is important that deep foundations penetrate through any loose, weathered rock that was exposed at the ground surface and weathered prior to being covered under younger deposits. Penetration can be aided by hardened steel tips attached at the bottom of a driven pile.

Limestone is vulnerable to chemical weathering that can create troublesome solution cavities, as shown in Fig. 1.7. The cavities often are clay-filled and hidden under more recent deposits, so unless detected by exploration drilling or geophysical methods they can come as an unpleasant surprise.

The most common igneous rock is granite, which forms the core of major mountain ranges. Granite also is the dominant "basement rock" that underlies continents, where it normally is buried under younger, sedimentary rocks. Granite normally is capable of supporting large foundation loads but foundations must extend down through loose or weathered layers.

Another common igneous rock is basalt, the black, solidified lava that emerged along plate margins. Its presence can be a signal for other potential problems such as volcanic activity.

Rock Quality

As shown in Fig. 1.1, granite can inherit cracks from mountain building, and cracks become open invitations for weathering. A useful measure of rock competence is the "Rock Quality Designation" or RQD, proposed in the 1960s by engineering geologist Prof. Don Deere (1922–2018) at the University of Illinois. The RQD is defined as the ratio between core lengths that are 4 in. (100 mm) or longer divided by the total length of a core run. The core diameter is *NX*, 2.125 in. (54 mm). An RQD >0.75 is regarded as "good." Continuous cores are required for laboratory measurements of unconfined compressive strength.

Rock Sockets

A rock socket is created when a boring is extended downward into rock in order to create resistance to lateral forces, for example, from cross-winds and currents on bridges. Deep foundations also may be used to resist lateral soil pressures associated with retaining walls.

Figure 11.6 shows lateral test data for a drilled pier with one-half of its total length embedded in rock. Reinforcing steel is concentrated near the perimeter of the pier. The maximum lateral force was over 80 tons, which is comparable to a vertical bearing capacity.

A critical component of a rock socket is the nature of the rock, as it must be strong enough to resist the lateral loading but also weak enough to permit drilling. Most of the cost usually goes to the drilling. Pressuremeter tests can establish a limit pressure in yielding rock. In harder rock a similar measurement can be made with a separation device called a Goodman Jack. In yielding soils acceptance also can be based on dilatometer test data.

End Bearing on Sand

The Terzaghi equation for shallow foundations, Eq. (7.1), often is used as a starting point for calculating deep foundation end-bearing capacities even though the failure geometries are quite different. In Fig. 7.2, slip surfaces for shallow foundations are assumed to extend up to the ground surface, which almost certainly does not occur

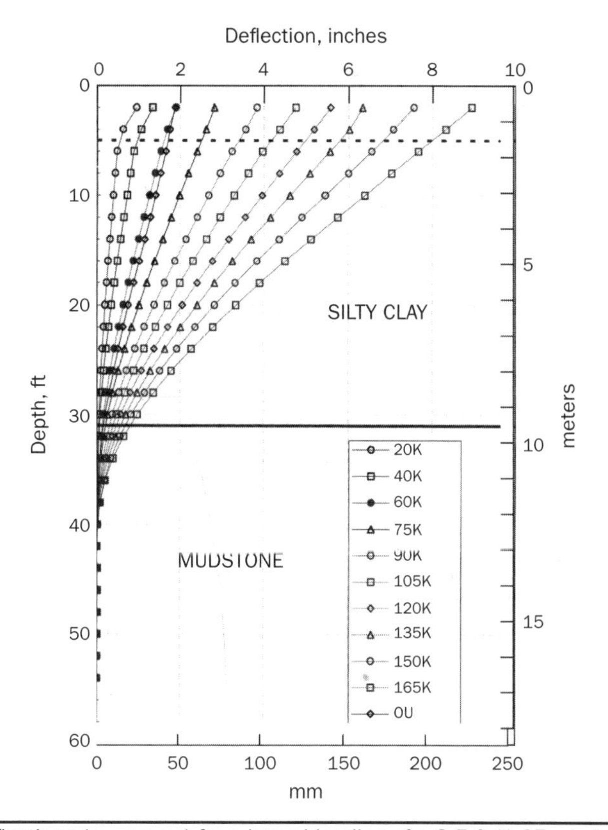

Figure 11.6 Deflections (*p-y* curves) from lateral loading of a 3.5 ft (1.07 m) diameter drilled concrete pier with the lower half socketed into mudstone, a hard shale without thin layering. 1K = 1,000 lb = 4.448 kN. (Graph source: Adapted from Nussairat et al., 2006.)

with deep foundations. Instead the failure geometry is more likely to resemble that created by penetration of a cone (Fig. 9.4). Another complication is because the classical derivations are two-dimensional, and punching failure of a deep foundation is three-dimensional.

The three dimensionless "*N*" factors in the Terzaghi equation are N_γ which is multiplied by a foundation width and is designated a width factor; N_c, which is multiplied by the soil cohesion and is a cohesion factor; and N_q, which is multiplied by depth and is a depth factor. Contributions from the width and cohesion factors to end bearing are small so they normally are ignored, which leaves:

$$q_t = \gamma' L N_q \tag{11.2}$$

where q_t is the end-bearing stress at failure, γ' is the soil unit weight on an effective stress basis, and L is the length of the foundation element. N_q is quite variable, and some selected trial values are as follows. The values for clay are included to demonstrate that its end bearing is relatively low.

Soil	Friction Angle	N_q
Plastic clay	0°	1
Soft clay	10°	2.7
Silt, loose sand	25°	13
Dense sand	35°	41

Example: A 1-ft diameter, 20-ft long pile is to be driven in loose sand with a unit weight of 100 lb/ft^3. What is the anticipated end bearing? What is a possible effect from submergence under water?

Answer: From Eq. (11.2), q_t = 100(20)(13) = 26,000 lb/ft^2. Multiplying times the end area gives $q_t = 26{,}000\ \pi\ r^2$ = 20,420 lb = 10 tons

Submerged, q_t = (100 – 62.4)(20)(13) = 9,776 lb/ft^2, a reduction of about 60 percent. A reduction of 50 percent for end bearing due to submergence is commonly assumed for sand.

A Critical Depth for End Bearing

Load test data summarized by Meyerhoff (1976) indicate little gain in end bearing in sand below a depth of about 10 to 20 pile diameters, perhaps related to the compressibility of sand at shallower depths.

11.13. Skin Friction and Adhesion

The deeper the foundation, the larger the surface area in contact with soil and the larger the amount of support that can be expected: (a) from side friction that depends on the normal stress, and (b) from adhesion that is independent of normal stress. The upper limit for adhesion is the cohesion of the soil.

Adhesion obviously is disrupted during installation processes of boring and pile driving, but then can slowly recover, which is a reason for delaying pile load tests. An even more demanding requirement that applies to cast-in-place concrete piers is to the time required for the concrete to gain strength, with 28 days being a common target. In practice, adhesion and friction along a deep foundation surface usually are lumped together and referred to as "skin friction."

Depth and Differential Movement

Differential movement is required to mobilize friction. Prof. Lyman Reese (1917–2009) and his coworkers at the University of Texas attached strain gauges to reinforcing steel in drilled piers, and discovered that side forces are not mobilized (1) in soil near the bottom of a drilled pier, where soil tends to move down with the pier, and (2) at shallow depths where drying shrinkage may cause soil to lose contact with the pier. The recommended lengths at both ends that are excluded from skin friction calculations are 5 ft (1.5 m).

Another complication can be that it is unlikely that end bearing and skin friction will peak out together. With increasing load, side friction will be mobilized the top down as the foundation element compresses, and most likely will peak out before end bearing is mobilized because of a factor of safety. This is suggested by the lightly shaded arrow at the bottom left in Fig. 11.7.

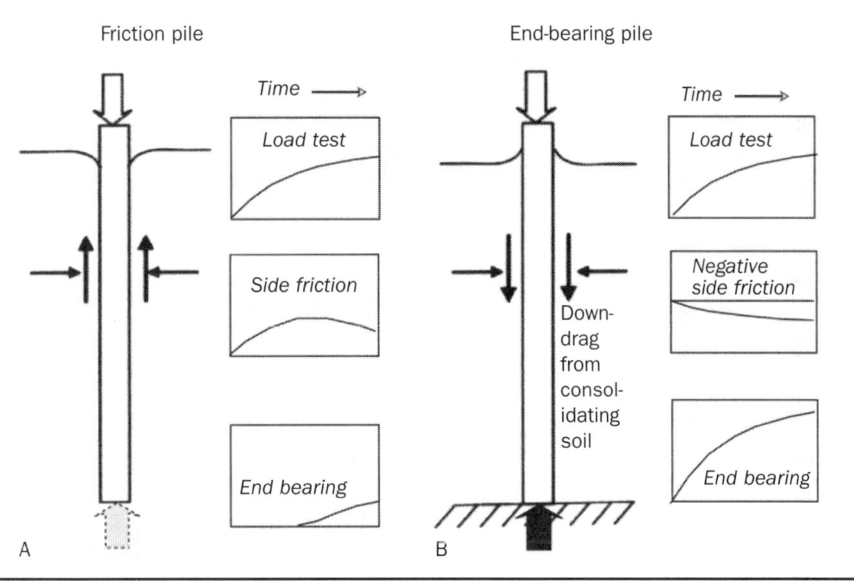

FIGURE 11.7 (A) Restraint develops and can peak out before end-bearing is mobilized. (B) A reduction in buoyant support from a lowered groundwater table can reverse the side forces so they push down and add weight that must be supported by a deep foundation.

Negative Skin Friction (Adhesion)

After a foundation is in service, lowering of a groundwater table can reduce buoyant support for the soil so it compresses and creates *negative skin friction* that pushes down on the foundation and subtracts from the bearing capacity. This is illustrated at the right in Fig. 11.7.

End Bearing and Skin "Friction"

As in the case of a cone test, skin "friction" on a pile or pier can be dominated by adhesion that is independent of normal stress, and is time-sensitive and influenced by thixotropic setting. Any frictional component is affected by excess pore water pressure that dissipates after driving stops. The full impact from aging on adhesion may not be realized for months or years after installation.

Uplift from Expansive Clay

Swelling of the upper, "active" layer of expansive clay can work to pull a deep foundation upward so it loses end bearing. This is most likely to occur with structures such as bridges where loading is heavy and transient. If the bridge itself is lifted, it can contribute to the "bump at the end of the bridge." Uplift may be minimized or prevented with a coating of soft bitumen or other material to prevent bonding between soil and foundation in the upper, "active zone" of the expansive clay. Another procedure involves the use of "bell-bottomed" caissons, discussed in the next section.

11.14. Drilled Shaft Foundations

Drilled shaft foundations were introduced in the early 1900s, coincident with the introduction of Portland cement. Borings initially were excavated by hand and often were performed inside pressurized enclosures called *caissons* to keep water out, and the name to some extent has carried over. Now power drills are used, and the more common designation is *drilled shaft*.

A short section of drill that has the same diameter as the finished shaft can be advanced down into the soil, then raised to the ground surface and the soil spun off. Soil exposed by repeated in-and-out movements of a drill ideally should be self-supporting. If it is not, a steel casing can be advanced with the boring.

After a boring reaches the full depth, a pre-assembled steel reinforcing "cage" is lowered and centered, and as the casing is being pulled, concrete is introduced from the bottom up through a pipe or a canvas tube called a *tremie*. The concrete is vibrated to prevent incorporating air pockets. As the fluid pressure from a column of concrete normally exceeds the lateral in situ stress in the soil, the column will tend to expand the outward and stabilize the contact pressure before the concrete sets.

A Bad Scene

Soil lacking cohesion may slough off and fall into a boring before it is filled with concrete, which can ruin a pier. Where there is even a remote possibility that this may occur, pier integrity can be examined with sound waves or diamond-bit coring.

Slow Demise of the Belled Caisson

A special cutter with hinged, swing-out blades can create a cone-shaped cavity at the bottom of a drilled pier and increase the area of end bearing. The enlarged "bell" also can serve as an anchor against uplift from expansive clays. However, installation requires manual downhole cleanout and inspection before concrete is poured, and working next to an unsupported ledge of soil can raise safety issues. The situation can be aggravated by groundwater problems, and the method has fallen into disuse.

11.15. Saving Time and Money on Load Tests with the Osterberg Cell

In 1984 Prof. Jorj Osterberg (1915–2008) at Northwestern University devised a method for load-testing drilled piers by pushing up from the bottom. The O-test® uses a large, expandable, sacrificial load cell that is inserted into the bottom of a boring prior to pouring concrete and inserting the reinforcing cage. After the concrete has set and cured, the cell is expanded with fluid pressure so it simultaneously pushes down on the soil and up on the pier, and the large cell area can convert fluid pressures into large forces. Failure occurs (a) when the shaft overcomes skin friction and moves up, or (b) soil under the cell suffers a bearing capacity failure and moves down, whichever comes first.

A test arrangement is illustrated at the left in Fig. 11.8. At the right in the figure, a cell is shown at a higher level to aid in mapping skin friction versus depth.

Representative Test Results

Figure 11.9 shows representative test results with an Osterberg cell at the bottom of a drilled pier. Settlement from compression of soil under the cell is shown in the bottom

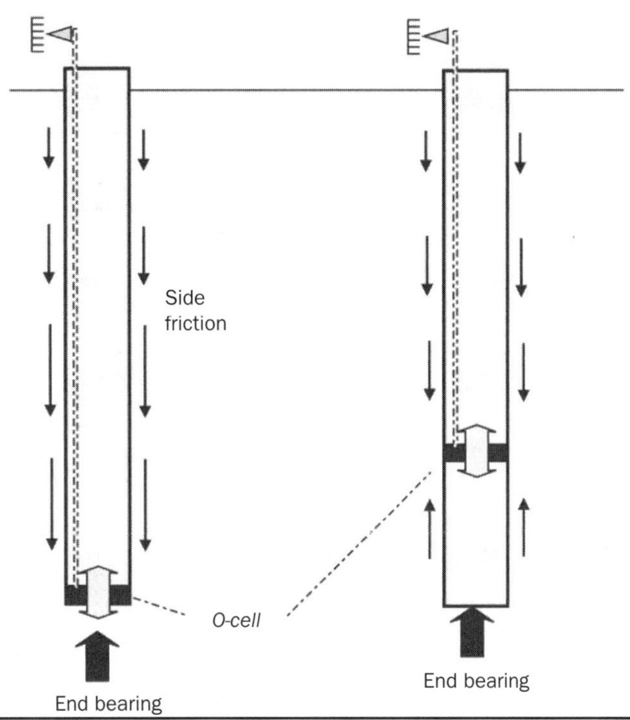

Figure 11.8 Two positions of Osterberg cells in drilled shaft piers to help to isolate influences from side friction. The usual arrangement is shown at the left, and the one at the right can be used to help isolate influences from skin friction. Telltale rods can monitor bottom movements and are isolated from the concrete.

graph, and upward movement of the pier is shown in the top graph, a distinction that is not made in a conventional top-down load test. The maximum applied load will depend on which fails first. As stress at the bottom may not be sufficient to cause plunge, a failure point may be defined on the basis of the sum of both movements.

Comparisons with Top-Down Load Tests

Comparable results between conventional and Osterberg cell load tests indicate that bottom support is essentially the same, but side shear develops more rapidly when mobilized from the bottom-up where the soil is likely to be denser and harder, and the contact normal stress is higher because of fluid pressure from the concrete.

Interpretations of Osterberg cell test data usually are provided by the firm conducting the tests, but the project foundation engineer should have a clear understanding of the test procedure and interpretation. Concrete shrinkage also can influence side friction but is minimized if drying is prevented.

Question: Which failed first in Fig. 11.9, end bearing or side friction?

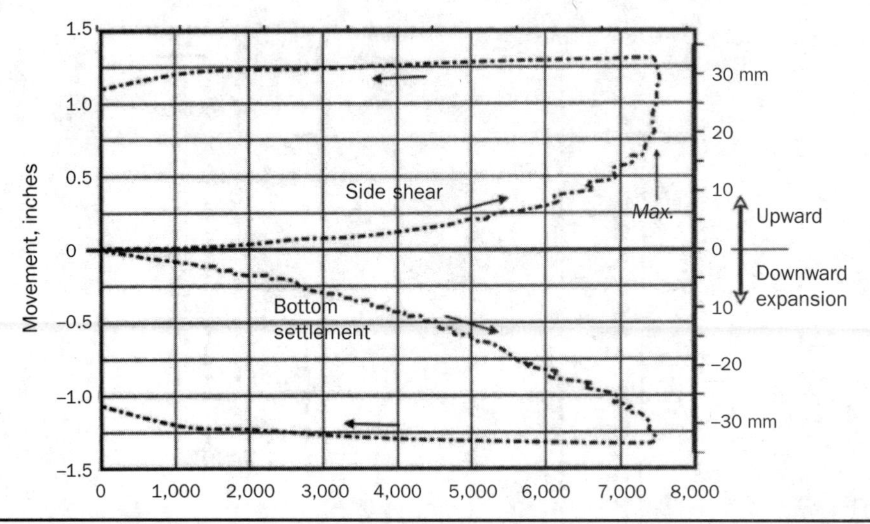

FIGURE 11.9 Deflections measured with an Osterberg cell during a load-unload cycle of a large, 9-ft (2.7 m) diameter concrete pier. Side shear failed first. (Data are from Osterberg.) (Graph source: *Geotechnical Engineering: Soil and Foundations Principles and Practice* by Richard L. Handy and Merlin G. Spangler. © 2007, McGraw-Hill Companies.)

11.16. Franki Piles

In 1909, Belgian engineer Edgard Franki devised a way to increase the bottom contact area of a concrete pier, by ramming a slug of concrete so it bulges out at the bottom. This is accomplished by lowering a steel casing into an open boring, dumping in a charge of zero-slump concrete, and mechanically ramming the concrete so that it pushes out to make a bulb. The noise from hammering is muffled by the casing depth. A reinforcing cage then is inserted and the casing normally is pulled as the remainder of the pier is poured. Franki piles are widely used, and are particularly adapted for use in loose sand. An obvious precaution is that positive fluid pressure must be maintained, so the casing must not be pulled too fast.

11.17. Augercast Piles

In the early 1960s, Norman Liver, an engineer with the Intrusion Prepakt Co., devised a novel method for holding borings open in loose or squeezing soils. The boring is made with a continuous hollow-stemmed, helical (screw-type) *flight auger* that is the full length of a pile. As shown in Fig. 11.10, the auger is suspended from a crane and rotated with a hydraulic drive motor at the top to make a boring. As the auger penetrates downward, excavated soil travels upward to fill the spiraling space on the auger and holds the boring open. After the full length is reached, grout is pumped down through the hollow auger and pushes out a sacrificial plug, then fills the boring as the auger is withdrawn. Positive fluid pressure is maintained during the withdrawal.

Steel reinforcing then can be lowered into the grouted column, and a cardboard tube is used to extend the pile to the design height. As fluid concrete normally has a

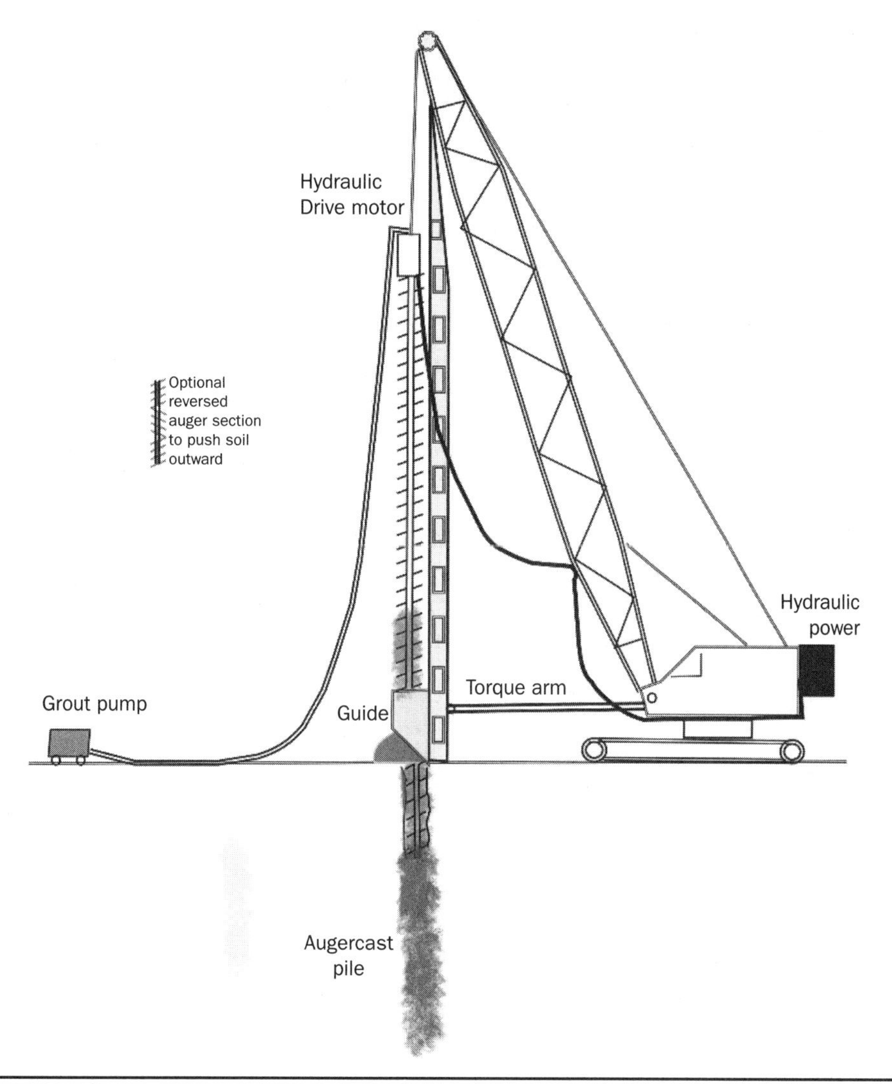

FIGURE 11.10 Installation of augercast piles. The auger goes to full depth and the boring grouted from the bottom up. High elevation of the grout pipe helps to maintain a high fluid pressure.

higher unit weight than the surrounding soil, the column can expand until it establishes an equilibrium condition.

An innovation is to incorporate a section of auger with a *reverse spiral* that can push grout outward to increase the diameter of the pile and reduce the amount of excavated soil that must be disposed.

Augercast pile diameters can vary from about 1 to 3 ft (0.3 to 0.9 m) and individual pile capacities usually are in a range of 40–200 tons. Installation is relatively quiet and can be performed close to existing structures. Piles also can be arranged in a line to create a foundation wall.

Jet-Grouted Micropiles

High-pressure jets can be used to cut away at soil while introducing a chemical hardening agent. Diameters usually are small, 4–7 in. (100–170 mm), and lengths can vary from about 20 to 80 ft (7 to 26 m).

11.18. Common Piles Materials

Steel pipe piles can have closed or open ends, and can be driven or vibrated into place. They are useful for penetrating through soft soil layers and achieve end bearing on a firmer stratum. The pipes then may be filled with grout.

Tapered piles are advantageous for maintaining and increasing contact pressure and side friction as they are driven.

H-beam piles are common and have a small displacement area. However, as a pile continues to penetrate, soil tends to be retained between the flanges. Hardened steel tips are often used to aid penetration through weathered rock to achieve solid end bearing. H beams also can be used as a friction pile, in which case about half of the side contact area is with soil that is held between the flanges.

Helical screw-in micropiles come in short sections, so they are particularly useful where space is restricted, as for underpinning existing structures. Sections are added as they are screwed down into the soil, and can be jacked against to correct localized settlement of a structure. Noise and ground vibrations are minimized. Because of the small diameter, the maximum capacity is about 40 tons.

Precast concrete piles are suitable for driving, particularly in granular soils, and unlike drilled piers require no waiting time for the concrete to set.

Prestressed concrete piles are similar to precast, but contain cables that are held in tension until the concrete has set. They may have a higher capacity and have a better resistance to bending moments.

Definitions of a Factor of Safety

As previously discussed, the factor of safety for a shallow foundation can be (1) based on the amount of load that it can carry, or (2) on the shearing resistance of the soil involved in a bearing capacity failure. The latter is more realistic, but it also is lower so it can give an impression of being less safe. It also is more difficult to calculate and seldom is used.

With deep foundations the two definitions merge, because most of the area of failure is defined by the length and diameter of the pile. The area involved in shear failure at the bottom is relatively small. The calculated factor of safety therefore automatically is lower for a deep foundation. A factor of safety of 3 still may be used but it often is reduced to 2, and some building codes allow a factor of 1.67 when based on full-scale load tests.

11.19. Preliminary Estimates for Deep Foundation Bearing Capacity

A preliminary estimate is required prior to making a load test, in order to have something to test. The estimate may be based on local experience, and/or inferred from in situ soil tests. The Standard Penetration Test (SPT) is conducted by hammering a sampler into the soil, more or less simulating driving of a pile. Cone tests involve

Remolding of soil during testing	3
Unknown soil lateral stress	3
Temporary excess pore water pressure*	2
Strength recovery through aging	2
Assumptions regarding a depth factor	2
Interpretation of data	1

*Can be detected with a piezocone.

TABLE 11.2 Some Factors Affecting the Use of Penetration Tests to Predict the Bearing Capacity of Deep Foundations on a Scale of 1 to 3, with 3 Being Judged to Be the Most Problematic

continuous pushing so it might be anticipated that the SPT may provide a better prediction of driving resistance, and the cone test a better measure of the final bearing capacity including end bearing and skin friction. Some factors that can affect the correlations are suggested in Table 11.2.

Based on analyses of published data, Meyerhof (1976) suggested preliminary estimates of deep foundation bearing capacity from SPT data. He also suggested a "critical depth" equal to 10 to 20 pile diameters, below which skin friction stays constant.

As long as end bearing does not encounter rock or hard soil, the contribution from skin friction can be dominant. Both the SPT and cone have tubular components that engage side resistance. In cone tests it is measured separately:

Soil	End Bearing		Skin Friction	
	Bored	Driven	Bored	Driven
Sand	≤1.2 N	≤4 N	$N_{av}/100$	$N_{av}/50$
Silt		≤3 N		

End Bearing in Clay

According to the Terzaghi bearing capacity equation, with $\phi = 0$, $N_\gamma = 0$ and $N_c = 5.7$. In foundation design it is the bearing capacity at the upper end of a deep foundation that is relevant, which eliminates the surcharge term N_q. In order to take into account the three-dimensional nature of a bearing capacity failure under a deep foundation, N_c is commonly boosted up to 9. Hence

$$q_t = 9c \tag{11.3}$$

where c is the soil cohesion. This relationship is commonly used, and the amount of error is diminished because side contact area for a deep foundation should far exceed the tip area and tip resistance. For a round pile or pier that is uniform in diameter, the tip resistance becomes

$$Q_t = 3.5D^2Q_o \tag{11.3a}$$

where Q_t is the total tip resistance, D is the diameter, and Q_o is the clay unconfined compressive strength with the friction angle ϕ assumed to be zero.

Skin Friction (Adhesion) between Drilled Piers and Clay

The cohesion of clay imposes an upper limit for bonding between a pier surface and clay. An arbitrary reduction factor $\beta = 0.5$ often is used for stiff clay. Bonding at the contact may be improved by cementation with lime liberated by hydrating Portland cement, but that only serves to move the shear surface outward and slightly increase the effective pier diameter. The total side restraint therefore equals βc times the side contact area of a drilled pier.

Load tests conducted by Reese at the University of Texas indicate that the total length L should be reduced about 5 ft (2 m) at the lower end where soil tends to move with the foundation, and at the upper end in expansive clay because in a dry season the clay is likely to lose contact with the foundation. The resulting equation for a round pier is

$$Q_s = \beta c \pi D L_e \tag{11.4}$$

where β is a reduction factor that varies from about 1 to 0.5 for soft to stiff clay, c is the soil cohesion, D is the pile or pier diameter, and L_e is the effective length of the foundation.

Example: In situ shear tests give $c = 1.2$ lb/in^2 (173 lb/ft^2) in clay that is penetrated by a 15 ft long, 2.5 ft diameter pier in expansive clay. Calculate a safe bearing capacity that can be attributed to skin friction.

Answer: This is not a stiff clay so $\beta = 1$, but the pier length L is compromised so $L_e = 5$ ft. Then $Q_s = (173 \text{ lb/ft}^2)(3.14)(2.5 \text{ ft})(5 \text{ ft}) = 6{,}800 \text{ lb}^2 = 3.4$ tons.

Skin Friction between Driven Piles and Clay

Bonding between pile-clay contact surfaces is complicated by (a) remolding of clay close to a pile surface, and (b) a potential for the developing high pore water pressure in saturated clay that is compressed during driving. Dissipation of excess pore water pressure can be relatively rapid and aid in creating a "setup factor" of 2 or more, which as previously discussed can put a halt to pile driving if it is interrupted.

Thixotropic setting requires more time, and can be evaluated by waiting a week or more before load testing. In sensitive clay it can push the setup factor to as high as 5 or more. An additional complication exists in layered systems where skin friction is an accumulation of contributions from different layers.

End Bearing in Sand

Contributions from width and cohesion bearing capacity factors to end bearing are relatively small so they often are ignored, which gives:

$$q_t = \gamma' L N_q \tag{11.4}$$

where q_t is the end-bearing stress at failure, γ' is the soil unit weight on an effective stress basis, and L is the length of the foundation element. However, the three-dimensional, confined failure geometry in soil near the bottom of a deep foundation contributes to wide variations in N_q that make accurate predictions difficult. That range in variability and comparisons to clay is indicated in Table 11.3.

Soil	Friction Angle	N_q
Plastic clay	0°	1
Soft clay	10°	2.7
Silt, loose sand	25°	13
Dense sand	35°	41

Table 11.3 Some Trial Values for N_q

Example: (a) What is the anticipated maximum end bearing for a 1 ft diameter, 20 ft long pile driven in loose sand with a unit weight of 100 lb/ft^3? (b) What may be the effect of submergence under a high groundwater table?

Answer: (a) According to Eq. (11.2), end bearing can be estimated as follows: The soil tip resistance is q_t = 100 lb/ft^3 (20 ft)(13) = 26,000 lb/ft^2. The end area is $A = \pi(0.5 \text{ ft})^2 = 0.785 \text{ ft}^2$. Then Q_t = 26,000 lb/ft^2(0.785 ft^2) = 20,410 lb = 10 tons. (The answer deserves no more than one significant figure.) (b) The soil unit weight is reduced to 100 – 62.4 = 37.6 lb/ft^3, which reduces the answer by (37.6/100) = 37.6%, or approximately 40%. A common approximation for the reduction of end bearing of a foundation on sand by flooding is 50 percent.

Lateral Stress "K" Values in Sand

Lateral stress is BIG in foundation engineering because it can dictate the amount of skin friction available. In cohesionless soils, lateral stress operates between certain limits that depend on the soil internal friction. Highest stress is developed in soil in a "passive state" that is resisting being pushed. An intermediate "at rest" state is inherited from a deposition process. Soil in an "active state" is acting against gravity to prevent sliding. The three states are entirely dependent on the soil internal friction, and cohesion must be zero.

Lateral stress only rarely is measured. Instead, it is inferred from other properties that can be measured. When lateral stress is measured, as in the next chapter, it leads to new insights regarding foundation bearing capacity. The relationships between *K*, the ratio of lateral to vertical stress, are summarized as follows:

Name	Occurrence	Symbol and Equation	
Passive	Resistance to being pushed	$K_p = (1 + \sin \phi)/(1 - \sin \phi)$	(11.3)
At-rest	As deposited	$K_o = (1 - \sin \phi)$	(11.4)
Active	Resistance to sliding	$K_a = 1/K_p = (1 - \sin \phi)/(1 + \sin \phi)$	(11.5)

Examples. With $\phi = 30°$, which is typical for sand:

1. Passive $K_p = 3.0$, three times the vertical stress. Example: Bulldozer.
2. At-rest $K_o = 0.5$, one-half of the vertical stress. Example: Natural soil.
3. Active $K_a = 0.33$, one-third of vertical stress. Example: Sand sliding down a slope.

Equations (11.3) and (11.5) in the table on the previous page can be derived with the aid of a Mohr circle. Equation (11.4), known as the Jáky equation, defines lateral stress in cohesionless soil that is partially supported between vertical shear surfaces that are presumed to be created during a deposition process. A simplified derivation is in Handy and Spangler (2007).

A lower value of K can occur but is not stable. It occurs in wind-deposited loess (silt), loosely deposited from winds and never saturated with water so it is held together with capillary forces (negative pore water pressure). When such a soil is saturated, it collapses.

$K = 1$ in a liquid, which includes concrete prior to set. Some representative passive and at-rest values for K that occur naturally in sand are as follows in Table 11.4:

Friction Angle				
	25°	30°	35°	40°
K_p	2.46	3.00	3.69	4.60
K_o	0.577	0.500	0.426	0.357

Table 11.4 Representative Minimum (K_o) and Maximum (K_p) Values for Sand

Example: Calculate the skin friction at a depth of 5 ft and at the contact with a driven (displacement) pile in sand with $\phi = 30°$ and a unit weight of 100 lb/ft^3.

Solution: Vertical stress is $\phi_v = 5(100) = 500$ lb/ft^2. Lateral stress is passive, so $K_p = 3.0$, and $3(500) = 1{,}500$ lb/ft^2.

Some Approximations

Table 11.5 presents some values that can be used for preliminary estimations of skin friction. The circumference of an H beam is two flange widths plus two web widths. The friction angle ϕ_d for the soil-to-steel surfaces may be assumed to be 0.2ϕ.

Example: Estimate skin friction (and adhesion) for a 2-ft diameter driven steel pipe pile passing through loose sand between depths of 12 and 18 ft. The unit weight of the overlying soil is assumed to be 100 lb/ft^3.

Solution: For loose sand use $K_o = 0.6$, so at an average depth of 15 ft, lateral stress is (0.6)(100)(15) = 900 lb/ft^2. For sand-steel contact use $\phi_d = 0.2(25) = 5°$, so tan 5 = 0.087. Cohesion and adhesion are zero, so total side restraint is (900)(0.087) = 78.3 lb/ft^2. The side area is $\pi(2)(6) = 37.7$ ft^2, and $F = 37.7(78.3) = 2{,}950$ lb = 1.5 tons. For total resistance skin friction (and adhesion) from other sections of the pile must be added to the estimated end bearing. The design load is the total divided by an acceptable factor of safety.

11.20. Pile Group Action

The smaller diameter of driven piles compared with drilled piers leads to the use in groups in order to achieve a required bearing capacity. This is illustrated in Fig. 11.11A, where overlapping normal stress fields should boost friction and aid coupling of the

Soil	Friction Angle	K_o	K_a	K_p	c_d
Plastic clay	0°	1	1	1	c
Hard clay	10°	0.8	0.7	1.4	0.5c
Silt, loose sand	25°	0.6	0.4	2.5	0
Dense sand	35°	0.4	0.3	3.7	0

Table 11.5 Data for Estimating Skin Friction (Plus Adhesion) for a Deep Foundation

piles. As the sequence of pile driving may push lateral stress fields off-center (not shown), a better driving sequence might be symmetrical, 1, 3, 2, 4.

Overlapping base areas and crowding of the perimeter of a pile group indicate that the group bearing capacity should be less than the sum of the parts. An appropriate *reduction factor* is difficult to measure because of large loads required to measure the bearing capacity of a large group of piles. In Fig. 11.11A, normal stresses are concentrated where stress fields from adjacent piles overlap, and with the aid of a rigid pile cap, a pile group will tend to act as a single mass.

Driven piles displace and can densify soil so it can approach a passive state and its strength is increased. This can act to compensate for the reduced footprint, and some tests in driven pile groups in sand have indicated that no reduction may be required for a group action. That almost certainly is not the case in clay soils. Where a large group of piles is involved, the bearing capacity of the entire block may be analyzed as a shallow foundation bearing capacity but with an arbitrary reduced depth as a conservative measure.

Pile Separation Distances

In Fig. 11.11B the diameter of the pile group is increased so the end-bearing area is increased. Bottom stress is less concentrated, which should improve the group factor. Test data from various sources on granular soils indicate a relationship between average group efficiency and pile spacing: *No reduction* with a pile spacing of 5*D*, 25 percent reduction with a spacing of 3*D*, and 50 percent with a spacing of 2*D*, where *D* is the diameter of the pile (Tuan, 2016). With a wider spacing the size and weight of the pile cap obviously must be increased.

Pile Group Action Formulas

Numerous formulas have been developed to account for the interactions of piles in groups but field testing and confirmation are difficult because of the large forces involved. One of the simplest formulas is the Feld Rule, which is based on a geometric relationship, that each pile can create a "shadow effect" on an adjacent pile that reduces its bearing capacity. Based on conventional pile spacings, Feld proposed a 1/16 reduction in bearing capacity for each adjacent pile, and made no distinction between sand and clay.

A general guide for efficiency is about 70 percent with a minimum spacing of 2.5 pile diameters. More detailed calculations are complex and are available on computer programs but still are difficult to confirm with field evidence. Physical models involving soils require that scaling down of dimensions requires an equal scaling up of unit weights. That can be accomplished by conducting tests in a large centrifuge. Other

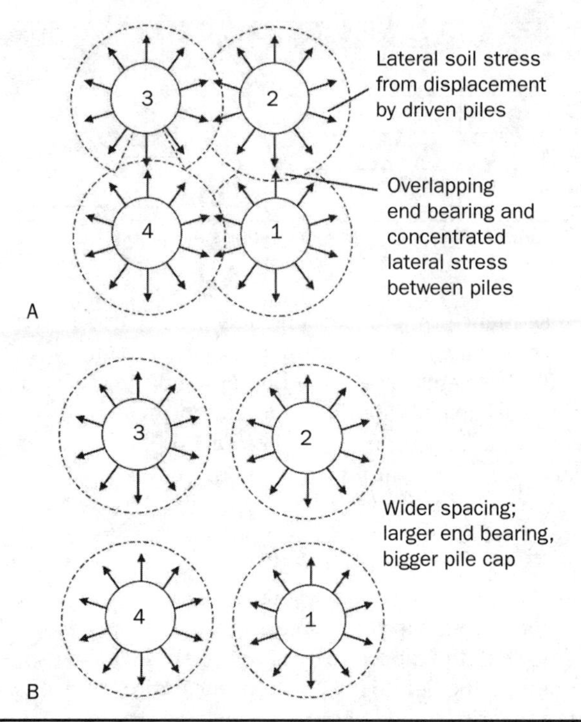

FIGURE 11. 11 (A) Overlapping stress fields from pile group action. (B) A wider spacing can improve efficiency.

options can include the use of a surface membrane and air pressure, or pumping water down through soil to create downward seepage forces. Additional insights into pile group behavior almost certainly will be investigated using instruments for measuring soil lateral in-sit stress such as discussed on p. 167ff in this book.

Batter Piles

The small diameters of batter piles compared to drilled piers make them more susceptible to lateral loads, so resistance to lateral loading can be accomplished by driving piles at an angle and connecting to vertical piles with a pile cap. The inclination angle can be determined by the angle of the resultant with a maximum anticipated horizontal load.

Questions

1. After driving a pile why should one wait before conducting a load test? After installing a drilled concrete pier how long should one wait?
2. An anchor pile is pulled upward and stops a load test. What conclusion can be drawn from the test? What are the options?
3. In Fig. 11.8 which is larger, end bearing or skin friction?

4. What deep foundation may be appropriate in soft, squeezing clay close to an existing structure? Is there a mechanism by which the amount of soil can be reduced that must be disposed of?
5. What are requirements for a deep foundation that is planned to engage a buried rock surface?
6. How may lowering the level of a groundwater table overstress a deep foundation?
7. Calculate end bearing and side restraint for a driven pile 1 foot in diameter 30 ft deep in clay soil with a unit weight of 100 pcf, an unconfined compressive strength of 500 psf, and a high groundwater table. What assumptions are necessary in the analysis?
8. Which is more effective for increasing skin friction in a granular soil, increasing the length or the diameter? Explain.
9. A "belled pier" that is enlarged at the bottom can serve as an anchor, and can increase end bearing in soil. What may be the downside?
10. Is bottom loading with an Osterberg cell likely to overpredict or underpredict settlement? Explain.

References

Day, R. W., *Foundation Engineering Handbook,* 2nd ed., McGraw-Hill, New York, 2010.

FHWA., *Design and Construction of Driven Pile Foundations: Lessons Learned on the Cenral Artery/Tunnel Project.* Project Publication No. FHWA-HRT-05-159, 2006. (Available online.)

Handy, R. L., and Spangler, M. G. *Geotechnical Engineering*, McGraw-Hill Companies, 2007.

Lambe, T. W., and Whitman, R. V., *Soil Mechanics,* John Wiley & Sons, New York, 1969.

Meyerhoff, G. G., "Bearing Capacity and Settlement of Pile Foundations," *ASCE Journal of the Geotechnical Engineering Division*, 102(GT3):197–228, 1976.

Nussairat, J., Liang, R., and Engel, R., *Design of Rock Socketed Drilled Shafts.* Report No. FHWA/OH-2006/21. U.S. Government Printing Office, 2006. (Available online.)

Tuan, P. A., "A Simplified Formula for Analysis of Group Efficiency of Piles in Granular Soil," *International Journal of Scientific and Engineering Research*, 7(7):15–21, July 2016. (Available online.)

Further Reading

Reese, L. C., and O'Neil, M. W., "New Design Method for Drilled Shafts from Common Soil and Rock Tests," *Foundation Engineering: Current Principles and Practices*, 2:1026–1053, 1989.

George Geoffrey Meyerhoff (1916–2003), professor and Dean at the Technical University of Nova Scotia, noted for his contributions to foundation engineering.

CHAPTER 12

Ground Improvement

Going Outside of the Box

12.1. What Is Ground Improvement?

Ground improvement refers to engineering applications and includes a multitude of different methods and approaches. It is an active area for innovation and research.

One of the oldest methods for ground improvement probably was discovered by accident, as straw incorporated into sun-dried bricks can reduce shrinkage cracking by promoting even drying. Geogrid is horizontally oriented plastic reinforcement that resists horizontal expansion and increases vertical strength. Reinforced earth acts in a similar manner but substitutes rows of parallel horizontal steel strips with ends fastened to a retaining wall. Tiebacks are long cables anchored in soil at one end and attached to retaining walls at the other. These methods are commonly used in geotechnical engineering.

12.2. Preloading

One way for a builder to skip the preliminaries is to preload a building site with a pressure that equals or exceeds those from anticipated foundation loads. Time must be allowed for settlement to occur, after which the preload can be removed and transferred to the next building site. The preload material usually is a pile of soil, and the method is particularly useful where a series of similar structures is planned. However, the method is not foolproof.

Preloading creates a "pressure bulb" illustrated in Figs. 5.2 and 5.4. The larger the contact area, the deeper the bulb and the more likely it is to extend below a groundwater table. Water expelled from consolidating soil is directed upward so a base layer of sand may be used underneath the preload to provide an exit for the water. Meanwhile the unit weight of the newly submerged soil is reduced about one-half, which slows and impedes the consolidation process.

Enhancing and Monitoring the Rate of Settlement

From a builder's point of view, time can be money so a drainage system may be justified. "Sand drains" are vertical borings filled with sand, and must connect at the top with a layer of sand to provide an exit for the water. An option is the use of plastic "wick drains" that are on large spools and are stitched down into soil in the manner of a sewing machine. Wick drains are hollow and perforated to allow entry of water.

Settlement induced by a surcharge load should be monitored to prevent premature removal of the surcharge. *Settlement plates* can be placed on the ground prior to placing the surcharge, and are connected to vertical pipes or rods that extend above the ground surface so settlement is obvious as the pipes appear to rise out of the ground. As they tend to be in the way of earth-moving equipment, another option is to set benchmarks on top of the embankment and monitor settlement with conventional surveying instruments.

A Complex System

As settlement normally slows down with time, it is conveniently graphed to a logarithmic time scale, as in a laboratory consolidation test. An acceptable stop time can be predicted with a first-order rate equation, discussed in Chap. 3. Delays caused by a rising groundwater table may not readily be recognized, but can be monitored using piezometers, small pipes with filters at lower ends so they act similarly to water wells. Water depths in the piezometers are measured with a tape that makes electrical contact with the surface of the water.

A surcharge load that tips or starts to sink into the ground is not a good sign, as it indicates shearing and a bearing-capacity failure. Particularly vulnerable are oxbow lake clays that are deposited and stay submerged under water so they are poorly consolidated. Shearing involved in a bearing-capacity failure can create smooth shear surfaces that permanently damage a clay soil.

Preloading should be approached with cautious optimism, and it is recommended that it be performed under the guidance of a geotechnical engineer.

12.3. Compaction

Establishing optimum moisture contents and acceptable soil densities for compaction is a routine operation in geotechnical engineering. Layer-on-layer compaction may be performed with tamping or "sheepsfoot" rollers that "walk out" to give a measure of the soil strength but only at the existing moisture content, which must be controlled.

Vibratory Compaction

Vibrating rollers are commonly used to compact soils for roads and foundations. Efficiency is increased if the vibration energy is such that the compactor lifts off and pounds the soil with each cycle. Lift-off becomes a major factor with hand-operated "jumping jacks" that are useful for compacting soil in restricted areas. Plate vibrators may use a slightly inclined vibration direction so they automatically inch forward.

Deep Dynamic Compaction (DDC)

An approach to site preparation that is more suitable in unpopulated areas is heavy hammering on the ground surface using a crane to repeatedly lift and drop a heavy weight (Fig. 4.12). Pounding is repeated and creates a waffle-like pattern of craters that can be meters deep and then must be filled using conventional methods. DDC is applicable for free-draining soils such as alluvial sand and gravel.

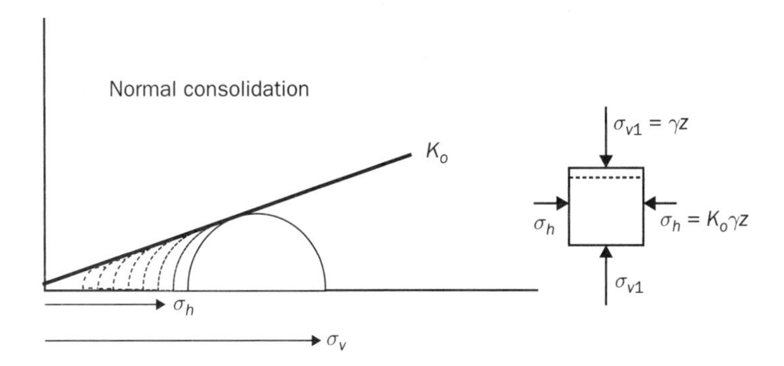

FIGURE 12.1 Mohr circle representation of vertical and lateral soil stresses during normal consolidation.

Blasting

Soil compaction with a row of explosive charges can be automatic if vibrations from one explosion charge sets off the next. Extreme diplomacy and caution are required.

Side Effects from Compaction

Soil that is compacted vertically will tend to expand horizontally so it can increase pressure on nearby walls. A light compactive effort therefore may be employed close to a basement wall. When used near a bridge abutment it can contribute to a familiar "bump at the end of the bridge."

Lateral stress imposed by a consolidating soil can be represented by a series of Mohr circles, as shown in Fig. 12.1. The ratio of horizontal to vertical stress is constant and is defined by K_o. It is high for clay and low for sand, so sand is a more suitable backfill. Least desirable is expansive clay that can be a common cause for inward buckling of basement walls.

12.4. Soil Replacement or Improvement

Some common methods used to strengthen soils are illustrated in Fig. 12.2. Their effectiveness is influenced in part by the kind of soil.

Stone Columns, Aggregate, and Mixed-in-Place Piers

Stone columns accompanied invention of the "Vibroflot" boring method in the 1930s in Germany. In this boring method, a vibrator incorporating high-pressure water jets is lowered into soil, so water cuts away and carries the soil to the ground surface. When the vibrator reaches a desired depth, coarse aggregate is poured into the boring and compacted while the vibrator is slowly pulled upward through the boring.

As crushed stone has very little cohesion, the vertical bearing capacity of a stone column mainly depends on lateral confining stress. A friction angle of about 36° has been measured in the columns, compared with that of sandy soil that usually is in a range of 25–30°. Bearing capacities of individual columns range from about 20 to 50 tons (189 to 450 kN), depending on diameter of the column and nature of the soil.

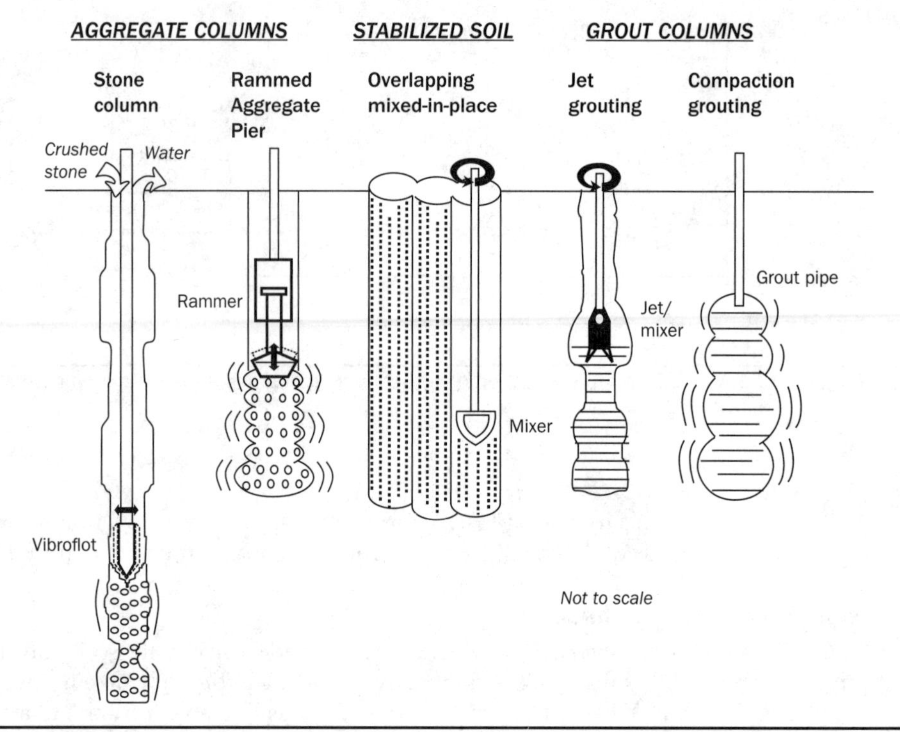

FIGURE 12.2 Some replacement methods used to improve the engineering behavior of foundation soils. (Image source: *Geotechnical Engineering: Soil and Foundations Principles and Practice* by Richard L. Handy and Merlin G. Spangler. © 2007, McGraw-Hill Companies.)

Stone columns have been used as area replacements that cut across active "pushing" shear zones in landslides.

Rammed Aggregate Piers®, or RAPS®, consist of columns of coarse aggregate hammered into place in layers. The method is increasingly used under the name Geopier®. The piers have been subjected to detailed studies of the influence of lateral soil stress, discussed latter in this chapter.

Mixed-in-place piers use a special drill to bore a hole and mix in a chemical stabilizer, usually hydrated lime or Portland cement. The drill is reversed so the mixture is compacted as the drill is slowly withdrawn from the boring. Overlapping mixed-in-place piers may be used to create walls without external forming.

Jet grouting creates a boring with jets of water or a cement-water mixture and fills the boring from the bottom up. The high injection pressure automatically enlarges the boring where it penetrates through less resistant soil layers.

Compaction grouting uses fluid pressure to create and expand a boring, which then is filled with a cement-water mixture. Lateral fluid pressure from the grout column acts to compact the soil but also requires drainage of the soil pore water so compaction grouting is most effective in permeable soils.

Injection grouting is commonly used to cement soil by injecting grout into cracks and rocks and soil voids. A Portland cement base is commonly used if soil pores are large enough to accommodate cement particles. Various *chemical grouts* can be used in finer-grained soils but increase the cost. One of the oldest methods uses two separate solutions, sodium silicate and calcium chloride, that react to create solid calcium silicate.

Ground freezing sometimes is used as an emergency measure to stem the flow of water.

12.5. Grout Materials

Most grout is a fluid mixture of water, Portland cement, and other chemicals that may delay the rate of setting and improve pumpability. A common additive is *fly ash* that is a powdery by-product from burning powdered coal. It is collected electrostatically from coal smoke, and mainly consists of tiny spheres of glass that improve fluidity. The ash then reacts with hydrated lime from the cement to create additional cementing compounds.

Cement grout is much denser than water, so grout in gravel may sink so it is not effective. Grout that invades an active seepage zone, such as in gravel or porous limestone underneath a dam, can be washed away before it sets. An important clue is if hydraulic pressure pushing the grout does not increase with time. Test borings may fail to detect any grout.

Curtain grouting consists of a line of grout holes so grout pressures tend to open cracks that link the holes and form a curtain. A common procedure is to use an initial wide spacing followed by half-spacings where increased resisting pressures can give measure of the effectiveness.

12.6. Grout "Take"

Grout pressure usually is monitored at the pump. A sudden drop in pressure can occur if the grout pressure is sufficient to open and invade cracks in rock or soil, and if pumping is continued can deliver it to far regions. That is called "take." Attempts to locate the grout with electrical resistivity or other in situ tests have had limited success.

Case history. High grout pressures were used to level a house on a slab-on-grade foundation on expansive clay. A transit was set up to measure movement. When the house did not move, the owner was advised to check inside to determine if any grout might be coming in. None was found until he tried to open a bathroom door.

12.7. Rammed Aggregate Piers

In the 1980s, an installation process now called Geopier® was developed by Dr. Nathaniel Fox. In this process, holes that typically are 30 inches (0.75 m) in diameter are bored and incrementally filled with crushed limestone aggregate. The aggregate is rammed into place in layers using a heavy hydraulic rammer. An initial goal was to save aggregate compared with cut-and-fill operations using construction equipment.

Piers were installed and load-tested, and used for small projects. Those successes led to the use to support a multi-story office building founded on loess soil in Memphis, Tennessee. The predicted maximum settlement was 1 inch (2.5 cm) based on pier load tests, but was so small *that it could not be measured*. As that violates all of the normal rules governing settlement, it merited further investigation. The rammer is shown in the cover photograph in this book, and is beveled to push aggregate outward and increase lateral soil stress. Might that somehow restrict the amount of settlement? Measurements of induced lateral in situ soil stress might be relevant.

Coincidentally, Dr. Fox was in-charge of research sponsored by the USDOT-FHWA and subcontracted to Iowa State University to develop a rapid, accurate method to measure lateral stress in soil. That research resulted in development of the K_o Stepped Blade, described in the preceding chapter in this book. Each data set requires about 15 minutes, and can provide several independent measurements at nearly the same depth, as shown in Fig. 10.6. Data are sufficient to allow calculating statistical means and standard deviations that are shown by points and short vertical lines in Fig. 12.3.

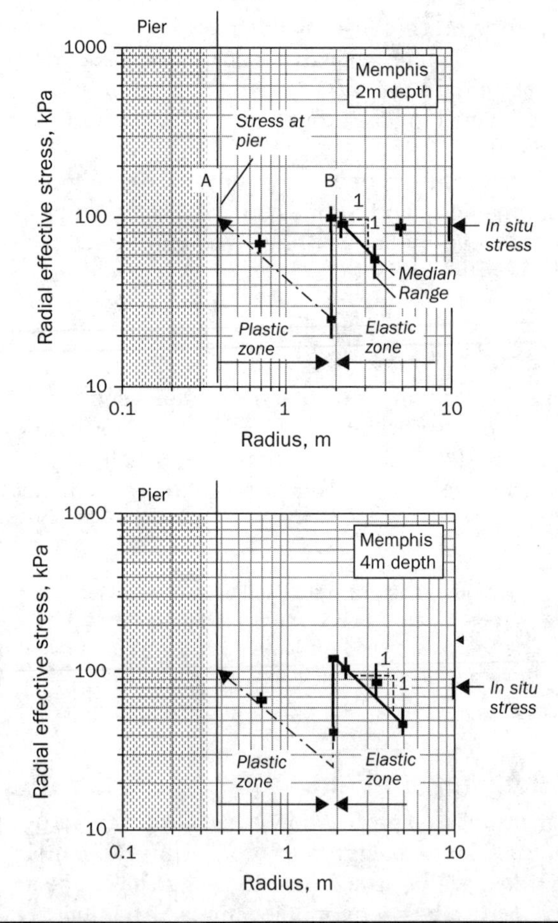

FIGURE 12.3 Results from lateral in situ stress measurements made with the K_o Stepped Blade in soil near a Rammed Aggregate Pier in overconsolidated silt (loess) soil. (From Handy, R. L., and White, D. J. "Stress Zones Near Displacement Piers." *Journal of Geotechnical and Geoenvironmental Engineering*, 132(GT1):54–63, 63–71, 2006. Reprinted with permission of the American Society of Civil Engineers.)

A "Saw-Tooth" Stress Pattern

The data in Fig. 12.3 were obtained from the same loess soil that supports the office building that did not settle. The radial stress declined linearly with increasing distance from the pier surface, as may be expected as the stress becomes distributed over an increasing perimeter area. Extrapolation back to the nominal pier surface gives a consistent pier-to-soil contact pressure of about 100 kPa (14.5 lb/in^2; 1 TSF). The high contact stress is consistent with the bevel around rammer edges.

Temporary Liquefaction

The transfer of lateral stress without loss, labeled *A* to *B* in the upper graph in Fig. 12.3, indicates a liquid behavior that in soil is designated by *liquefaction*. It occurs in loose granular soils as grains are pushed into adjacent water-filled voids.

The peak-to-peak *A-B* distance in the figure indicates that liquefaction extended to a radius of about 1.9 m (6 ft). This interpretation is indicated in the shaded area in Fig. 12.4. It may be noted that the same process of temporary liquefaction probably aids pile driving in saturated, fine-grained soils.

Outside of Point B, radial stress decreased linearly, which is expected if it is distributed around an increasing perimeter. (The 45° slope of the line is the same with both linear and logarithmic scales.)

Tension Cracks Outside the Liquefied Zone

According to elastic theory an increase in radial stress outside of the liquefied zone should be accompanied by a decrease in tangential stress. Soil, being weak in tension, should be expected to develop radial tension cracks such as shown in Fig. 12.4. Crack lengths will be influenced by the pre-existing soil K_o stress that increases with depth, the degree of overconsolidation, and the presence and hydration state of expansive clay minerals.

Crack lengths are not well defined by 2-m depth data in Fig. 12.3, but at 4-m depth appears to extend to a radial distance of about 5 m (16 ft), several meters outside of the

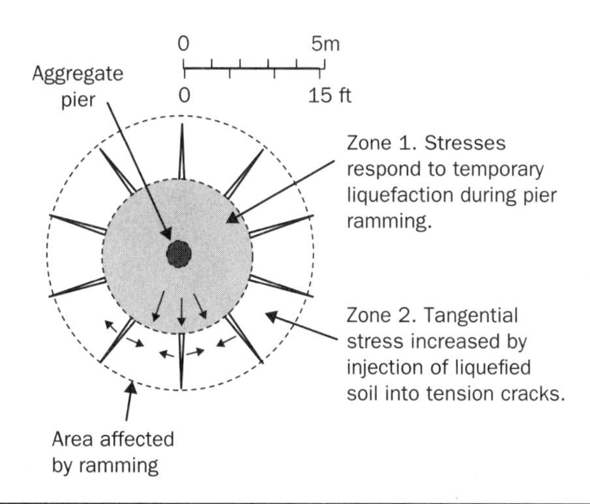

FIGURE 12.4 Zones of temporary liquefaction and radial tension cracking. Stress conditions interpreted from data in Fig. 12.3.

zone of K_o compressive stress that existed in the soil. This attests to a high fluid pressure created by ramming of liquefied soil into the cracks, and elastic stress concentration factors at ends of the crack. At 4-m depth, the area of soil invaded by tension cracks is about 40 times the cross-sectional area of the pier prior to ramming.

12.8. A Hypothesis of Friction Reversal

Figure 12.5 shows a sequence of diagrams showing hypothesized friction reversals at soil grain contacts as a consequence of lateral ramming. In theory the direction of friction can be reversed without significant sliding.

Figure 12.5*a* depicts shear stresses at grain contacts in normally consolidated soil. Intergranular friction acts to partially support the soil so lateral stress (horizontal black arrows) is less than vertical stress from weight of the soil.

In Figure 12.5*b*, friction arrows are reversed by lateral ramming pressure so horizontal stress exceeds vertical stress from weight of the soil.

Figure 12.5*c* illustrates how a vertical foundation load then must be sufficient to *again reverse the friction at grain contacts* before sliding can occur and allow significant settlement.

This influence should extend beyond the perimeter of a pier because of the nature of the pressure bulb.

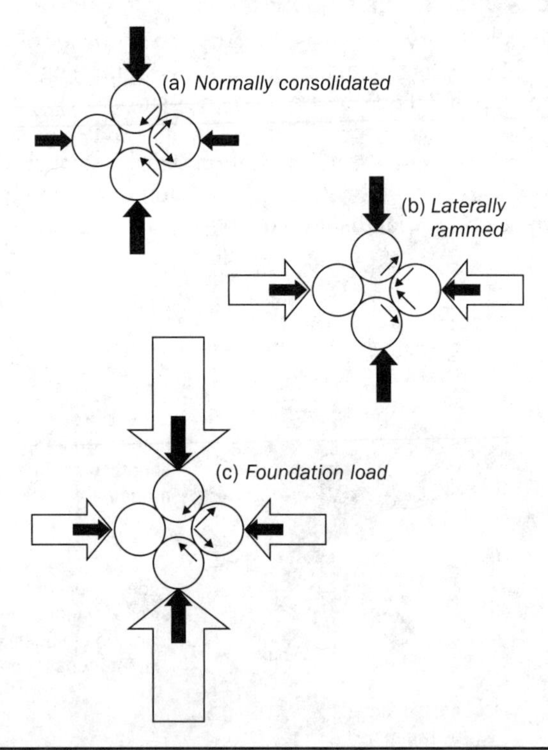

FIGURE 12.5 Hypothesis of friction reversal to reduce or prevent foundation settlement: (a) lateral pressure from ramming reverses friction directions at grain contacts; (b) contact friction must again be reversed through foundation loading, (c) before settlement can begin.

Conditioning

By definition there can be no friction at the grain contacts in liquefied soil. A friction reversal therefore cannot occur until after the liquefied state passes. This probably is contributed to by the incremental aggregate deposition and ramming, which may allow time for partial drainage and a reduction in the pore water pressure. The reversal of friction then may occur as a result of ramming pressures on successive layers of aggregate.

In summary, friction reversals at grain contacts appears to be consistent with essentially zero foundation settlement, and can be accomplished with the ramming procedure used to install and compact Rammed Aggregate Piers.

Friction Reversal and Overconsolidation

Removal of overburden pressure reduces or takes away the vertical arrows in Fig. 12.5*a*, which should reverse friction at grain contacts (Fig. 12.5*b*). Friction again must be reversed by vertical loading before consolidation can begin. *Lateral stress created by ramming therefore appears to mimic overconsolidation that results in an increase in the soil shear strength.* Aging at grain contacts should aid in preserving the strength gains.

12.9. Advanced Course: Application of Mohr's Theory

Figure 12.6*a* depicts a progression of Mohr circles in a horizontal plane in a cohesionless sand during pier ramming: As radial stress σ_r increases, tangential stress σ_t decreases, until it engages a failure envelope and the soil shears.

MOHR CIRCLES IN A HORIZONTAL PLANE

(a) Rammed boring in cohesionless soil:
Shear failure, no tension cracks

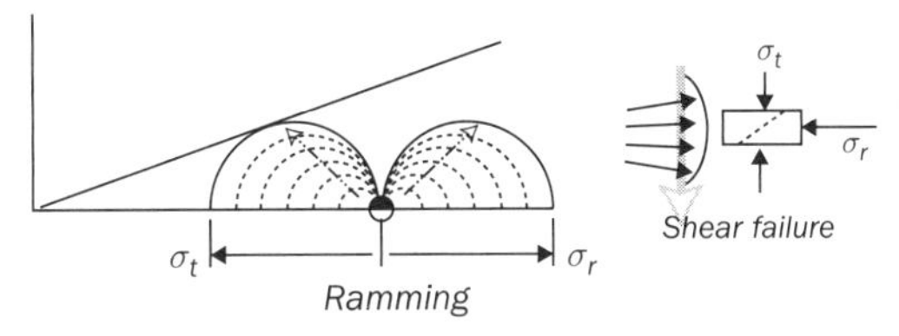

(b) Rammed boring in saturated cohesive soil:
Tension cracks form and are extended with liquefied soil

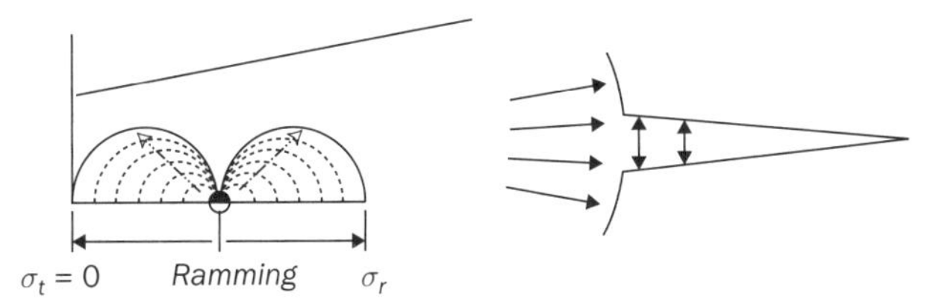

FIGURE 12.6 (a) A cohesionless soil structure may shear and collapse under lateral ramming pressure. (b) Tension can develop in a soil that resists compression, contributing to tension cracks that are extended outward by fluid pressure from liquefied soil until the pressure dissipates.

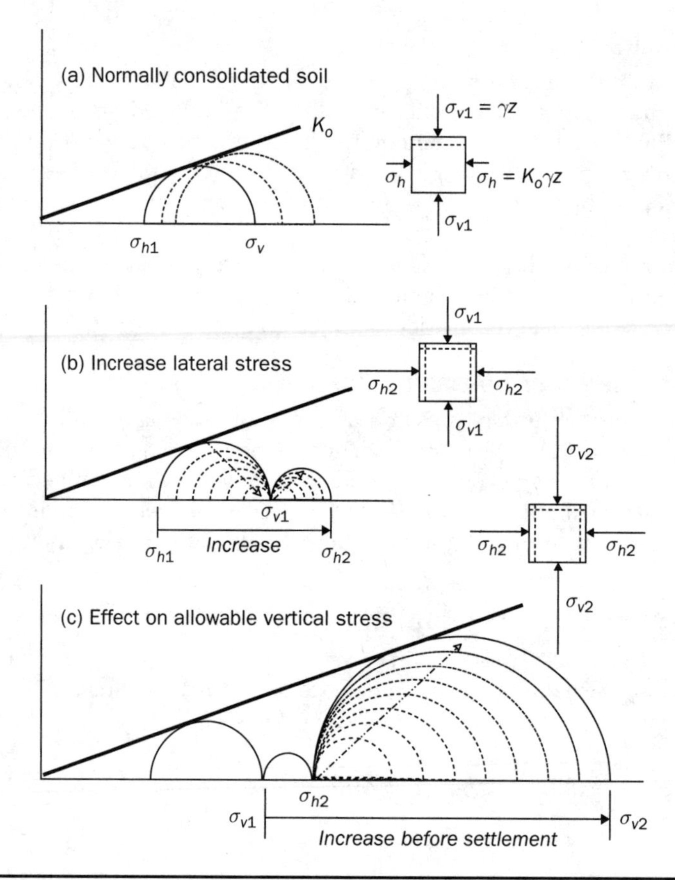

FIGURE 12.7 A two-dimensional explanation for the effectiveness of ramming: (a) Consolidation of accumulating soil shifts Mohr circles to the right. (b) Ramming imposes horizontal stress that often will exceed the vertical stress. (c) Vertical applied pressure then must be substantially increased before consolidation and settlement can occur.

Figure 12.6*b* is more appropriate for a cohesive soil: As radial stress σ_r increases, tangential stress σ_t decreases until it becomes tensile and creates tension cracks. If the soil is saturated and liquefied as a consequence of compression, it will squeeze radially outward into the cracks. Fluid pressure can aid in extending the cracks until viscous resistance saps away the expansion pressure. The maximum crack length therefore may not be readily predictable.

Lateral Stress and Settlement

Figure 12.7*a* represents vertical and horizontal stresses in normally consolidated soil. Figure 12.7*b* illustrates that vertical stress from ramming increases horizontal stress. If it exceeds vertical stress, friction will be reversed at grain contacts. A large vertical stress then will be required to again reverse contact friction so that consolidation can begin (Fig. 12.7*c*).

The process depicted in Fig. 12.7 represents stresses in a vertical, *xz*, plane and does not take into consideration stress relief that can occur in the *y* direction (normal to the paper). The analysis may best describe conditions in soil close to a line of piers than close to a single pier where lateral stress will be partially relieved in the y direction. Load tests on an isolated pier therefore may tend to overpredict settlement when there is a line or group of piers.

Is Excavation Permitted Close to RAPS?

Excavation will relieve lateral stress that contributes to soil strength, whereas aging can improve bonding at the grain contacts. Caution is advised, and an arbitrary minimum distance of 5 ft (1.5 m) may be used. Careful excavation may expose soil-filled vertical tension cracks.

12.10. Further Developments

RAPS as Anchor Piers

A steel plate can be placed in the bottom of a pier boring and connected to two steel rods welded to opposite sides of the plate and extending above the ground surface. After the pier is constructed, its capability as an anchor can be evaluated by pulling on the rods.

When Soil Does Not Hold an Open Boring

A boring can be made in soft soils by driving a thick-walled steel pipe. The pipe is filled with coarse aggregate that is retained by short lengths of chain that are welded on the inside, hand down inside the pipe, and jam up during ramming. After a layer is rammed, the pipe is pulled upward so the chains fall loose and release aggregate for the next lift, which then is rammed into place. The process is repeated until the pier is completed.

Low-Slump Concrete Piers

Concrete can be substituted for coarse aggregate and rammed into place in layers. The low water-cement ratio improves concrete strength, and lateral confining pressure improves skin friction. Advantages can be evaluated from comparative load tests.

Sand Piers

Experimental rammed sand piers can demonstrate the advantages from ramming. Coarse aggregate still is preferred as it develops a higher angle of internal friction.

Questions

1. Two piles that have the same lengths and diameters are installed in the same stiff clay soil. One pile is driven, the other augercast. Do you expect the driven or the augercast pile to have a higher bearing capacity? Might there be a relationship to the kind of soil?
2. What two steps be involved in a reversal of intergranular contact friction during ramming of liquefied soil Rammed Aggregate Piers?
3. It may be assumed that temporary liquefaction may aid pile driving. How might this be reflected in a setup factor?

4. What can go wrong with preloading a building site prior to construction? What physical tests may be appropriate?
5. Pile group action is mainly a matter of conjecture because load capacities are too large to be conveniently measured. How might measurements of lateral in situ stress aid in these evaluations. Suggest a test program to compare bearing capacities of a four-pile and a five-pile group with one pile in the center.
6. May friction reversal be affected by the kind of soil?
7. High ramming stresses are used during the installation of Franki concrete piles, discussed in Sec. 10.5. How does that process differ from the Geopier method?

Reference

Handy, R. L., and White, D. J. "Stress Zones Near Displacement Piers." *Journal of Geotechnical and Geoenvironmental Engineering,* 132(GT1):54–63, 63–71, 2006.

APPENDIX

The Engineering Report and Legal Issues

The Written Report

Nature is complicated, and foundation engineers must deal with the complications. Most soils are recycled from rocks and older soils, and every building site has a story to tell. As soils are highly variable, design is considerably on the safe side. Results from an investigation normally require a written report so the engineer who disdains writing skills may experience a legal equivalent of quicksand.

Organization of a Report

(a) **Introduction.** An introduction should briefly state the purpose of the investigation, and may cite obvious site limitations: Rivers cut away at their banks, floodplains flood, hillsides may slip away after prolonged rains, underground mines and caverns can collapse, earthquakes shake, and building sites can be complicated.

(b) **Site geology and pedology.** These are the keys that open the door: Geology defines the rock that weathers into soils, and geological soil deposits derive from activities of wind, water, and ice. Pedology involves processes of weathering and can fill in additional details. This section can be golden and often can be used in more than one report. It should begin with a "walkabout."

(c) **Site exploration methods and tests, goals and purposes.** Drilling and testing are limited to specific locations, where a goal is to identify the soils and measure relevant engineering properties.

(d) **Test results and interpretations.** This main presentation incorporates appropriate data, graphs, cross-sections and photographs.

(e) **Analysis, conclusions and recommendations.** The most important section that draws on all of the above.

(f) **Essential qualifier**: The information, conclusions and recommendations in a report *cannot guarantee what is between the sampling and testing sites*; hence the common requirement of a generous *factor of safety*. This section often is written with the aid of legal counsel.

(g) **In-house review**. A report often will be critiqued by a senior member of a firm.

(h) **Submission, with possible oral presentation and discussions.**

Legal Issues

Item: The expert witness must be knowledgeable and able to explain difficult concepts so they are interesting and understandable by a judge and jury.

Item: The expert witness is under oath to be truthful, so words must be carefully selected and honesty not only is the best policy, it is the *only* policy.

Item: Test borings can only reveal what it there so everything in between is an educated guess. Interpretations can be aided but not substituted by supplemental tests such as with echo sounding and ground-penetrating radar.

Item: A legal requirement for design and construction is that they must have been in compliance with the *state of the art that existed at the time that they were performed.* For example, the grain elevator that flopped over in Fig. 5.8 was built prior to an understanding of *effective stress,* the effect of pore water pressure on soil shear strength.

Item: *Failures can perform a valuable function* if they do not conform to the existing technology. For example, a retaining wall failure that involved multiple fatalities was not consistent with conventional design methods and analyses, and led to an analytical solution for arching action in soils.

Item: *Nothing in writing?* In order to gain some insight on how best to proceed, an attorney may request a site visit by an expert with nothing in writing that could be subpoenaed for trial. The expert still may be called by either side, but testimony based only on a site visit will be of limited value.

Item: *Expert witnesses working together?* Clients may suggest that experts on both sides work together to find a solution, in which case a difficulty may be that they not agree on a solution. Normally both sides have experts who do independent investigations and make their cases, and leave a final decision up to a judge or jury. It does not always work.

Item: *Engineers on a jury.* Some trial attorneys may request that engineers be excused from jury duty, because engineers are "too rational."

Meeting the challenge: ASFE

Although foundation failures are rare, in the 1960's insurance companies made note of liability issues and raised their rates. A crisis ensued, and in 1969 a large group of practicing engineers formed ASFE, the *Associated Soil and Foundation Engineers.* (The name has since been changed to the Geoprofessional Business Association).

An initial approach was to point out unavoidable limitations of engineering reports and try and make then more lawsuit-proof. Lawyers were consulted, and member educational programs initiated. Arrangements were made for member's reports to be impartially reviewed by other members and attorneys, to look for legal issues and point out vulnerabilities. A members-only insurance company was formed, and the program was so successful that the company lowered its rates and made refunds. Almost 100 case histories have been made available to members and can bring the real world into the classroom. In some university programs referrals from local consulting firms are used as class projects, and reports are made to the firms and the owners.

Index

Page numbers followed by "f" refer to figures and "t" refer to tables.

A

AASHTO (American Association of State Highway and Transportation Officials) system, 29f, 30–31, 31f
Abandoned mine shafts and tunnels, 20
Active layer (in expansive clay), 6
Active state soil pressure, 17
Additions, building, 55–56
Adhesion theory for sliding friction, 43
Agricultural soil maps, 9
  soil series, 9
A horizon topsoil, 60, 109
Air photo interpretation, 10f–11f, 12–13
Alluvial soils, distinguishing between, 9–13, 10f–11f
  air photo interpretation, 10f–11f, 12–13
  alluvial fans, 12, 71
  meanders and cutoffs, 10–11, 10f
  natural levees, 12
  oxbow lake clay, 10f, 11–12
  rivers and continental glaciation, 10
  slack-water (backswamp) floodplain deposits, 12
Alluvium, collapsible, 71
American Association of State Highway and Transportation Officials (AASHTO) system, 29f, 30–31, 31f
Amontons, Guillaume, 93
Amontons' First Law, 93
Amontons' Second Law, 94–95
Angle of repose, 13
Arches, 41
Arctic permafrost, 64, 121
Army Corps of Engineers, 137
Atterberg, Albert, 25, 40f
Atterberg limits, 17, 25
  and criteria for expansion, 31, 32f
  in soil classification, 29–30, 29f
Augercast piles, 8, 182–184, 183f
  jet-grouted micropiles, 184
Automatic hammer, 131f, 132t

B

Batter piles, 190
Bauxite, 60
Beaded streams, 121
Bearing capacity and settlement, 93, 94f
Belled caisson, slow demise of, 180
Bentonite, 34, 63
Between-layer stacking of water molecules, 68–69
B horizon subsoil, 60, 109
Bimodal, 35, 38
Black cotton soil, 6, 65, 66f
Blasting, 195
Blow counts, 126, 126f, 130
Borehole Shear Test (BST), 98–100, 98f–99f
Boussinesq, J.V., 79, 91f
Boussinesq equation, 79–81, 79f
Boussinesq theory, 80–81, 80f
Braided rivers, 11f, 13, 61
Brooming, 170
BST (Borehole Shear Test), 98–100, 98f–99f

C

Caisson, 180
California Bearing Ratio, 149
Carlson, L., 72
Casagrande, Arthur, 27, 30, 49, 58f
Casagrande liquid limit device, 27, 27f
Casagrande method, 53, 53f
Castles, 41
"Cathead," 130, 131f
Cathedrals, 41
"Cave-in-Restaurant," 114f
Caverns, 2–3, 18
  natural history of, 120
Chainsaw method, in managing expansive clays, 65
Chemical soil stabilization, 3
Chemical stabilization, 17
Chen, F.H., 31, 32f
Chen criteria, 71f
Chu-Davidson air-jet dispenser, 37f
Cities and rivers, 168
Citizen complaints, 175
Clathrate, 121
Clay
  end bearing in, 185
  foundation bearing capacities based on SPT or unconfined compressive strength, 136–137, 138f
  kinds of clay materials, 31–34
    drilling mud, 34
    expansive crystal structure, 32–34, 33f
    going tribal, 34
    layered crystal structure, 31–32, 33f
    when sodium (Na^+) replaces calcium (Ca^{++}), 34
  size grades, defining, 39
Claystone, 60
Coefficient of consolidation (C_v), 47–48
  evaluating, 49
Cohesion, soil, 143
Cohesive Nonexpansive Clay (CNC), 69
Cohesive shear strength, 143
Collapsible soils, 70–71
  collapsible alluvium, 71
  delayed collapse, 70–71
Columbia Plateau (United States), 63
Column foundations, 7
Compaction, 194–195
  blasting, 195
  deep dynamic compaction (DDC), 194
  side effects from compaction, 195, 195f
  vibratory compaction, 194
Compaction grouting, 196, 196f
Compression modulus, 152
Computers, 76
Concentrated stresses, 79, 79f
Concrete, 1
Conditioning, 201
Cone tests. *See* Probing with cone penetration tests and the Marchetti Dilatometer
Confidence limits, 104
Consolidation rate, 46–48, 47f
  defining a drainage distance, 48
Consolidation test, 59, 59f
Consolidometer, 42, 42f
Constrained deformation modulus, 152
Construction materials, three most common, 1–2
Continental drift, 21
Coriolis, G.G., 61
Coriolis effect, 61
Coulomb, Charles, 95
Coulomb's equation, 95–96, 97f, 124
County soil maps, 9
Crystal structure in control of expansive clays, 63, 64f
Curtain grouting, 197
Cut-and-fill, 109, 109f
C_v (coefficient of consolidation), 47–48
  evaluating, 49

D

Darcy's Law, 46
da Vinci, Leonardo, 27, 27f, 93
DDC (Deep Dynamic Compaction), 3, 74, 74f, 194
Deccan Plateau (India), 63
Deep, definition of, 2
Deep Dynamic Compaction (DDC), 3, 74, 74f, 194
Deep foundations, design of, 167–192
  analyzing hammer blows, 173–175
    measuring setup with PDA and restrike, 174–175, 175f
    pile driving analyzer (PDA), 173–174
    wave equation for driven piles, 173, 174f
  augercast piles, 182–184, 183f
    jet-grouted micropiles, 184
  citizen complaints, 175
  common pile materials, 184
    definitions of a factor of safety, 184
  drilled shaft foundations, 180
    bad scene, 180
    slow demise of the belled caisson, 180

Deep foundations, design of (*Cont.*):
end bearing, 169
Engineering News formula, 170–171
Franki piles, 182
lowering of sea level, 168
overview, 2, 8, 167
pile bearing capacities and load tests, 171–173, 171t
anchor requirements, 171, 172f
conduct of a test, 172
criteria for failure, 172, 173f
marginal designs, 172
strength gains and slow loading, 171
pile driving, 169–170
hard driving and brooming of wood piles, 170
no lunch breaks!, 170
science of hammering, 169–170
wood piles, 169
pile group action, 188–190
batter piles, 190
pile group action formulas, 189–190
pile separation distances, 189, 190f
pile load capacities: end bearing, 175–178
critical depth for end bearing, 178
end bearing on rock, 175–176
end bearing on sand, 176–178
rock quality, 176
rock sockets, 176, 177f
preliminary estimates for deep foundation bearing capacity, 184–188, 185t, 187t
approximations, 188, 189t
end bearing in clay, 185
end bearing in sand, 186–187, 187t
lateral stress *K* values in sand, 187–188, 188t
skin friction (adhesion) between drilled piers and clay, 186
skin friction between driven piles and clay, 186
saving time and money on load tests with the Osterberg cell, 180–181, 181f
comparisons with top-down load tests, 181
representative test results, 180–181, 182f
skin friction and adhesion, 178–179
depth and differential movement, 178, 179f
end bearing and skin "friction," 179
negative skin friction (adhesion), 179, 179f
uplift from expansive clay, 179
soils and city planning, 167–168
cities and rivers, 168
Deep foundations, design of (*Cont.*):
tension breaks in concrete piles caused by pile driving?, 170
piles doing a U-turn, 170
transferring a foundation load deep to reduce settlement, 167, 168f
when pile foundations became a matter of necessity, 167
Deep tropical weathering and expansive clay, 63
Deere, Don, 176
Delayed collapse, 70–71
Dependent variable, 104
Depth and differential movement, 178, 179f
Differential settlement, 55–56, 55f
problems with building additions, 55–56
Dilatancy, 96
definition of, 143
Dilatometer modulus, 152
Direct shear test, 96–100, 97f
Borehole Shear Test (BST), 98–100, 98f–99f
influence of layering, 97–98
DOT-FHWA (U.S. Department of Transportation, Federal Highway Administration), 161, 198
Doughnut hammer, 130, 131f, 132t
Drainage, 16
Drifts (mines), 119
Drilled quicklime, 17–18
Drilled shaft foundations, 180
bad scene, 180
slow demise of the belled caisson, 180
Drilled shafts, 180
Drilling mud, 34
Driving versus pushing, 142, 143f
"Drunken forest," 121
Dry weight, 25, 45
Dune tract, 13
Dutch cone, 141, 142f
Dynamic compaction, 3

E

Earthquakes, 21–23
buildings and, 23
conflicting possibilities, 23
dynamics, 75–76, 75f
with fracking, Oklahoma leads the nation in earthquakes, 74, 146
longest, most intense period of ground shaking recorded in the U.S. (New Madrid, Missouri), 73
man-made, 22, 74
recurrence intervals, 22–23, 75–76

Eccentric loading, 117–119
foundations for retaining walls, 117–119, 118f
Effective stress, 42–43, 42f
Eichner, Don, 162
Electrical resistivity measurements, 20
Elongated lakes, 121
End bearing, 8, 169, 175–178
in clay, 185
critical depth, 178
on rock, 2–3, 175–176
rock quality, 176
rock sockets, 176, 177f
in sand, 176–178, 186–187, 187t
skin "friction" and, 179
Engineered soil moisture content, 25–26
Engineering News, 170
Engineering News formula, 170–171
Engineering News Record, 170
Engineering seismograph, 20
Eolian silt deposits, 13–14, 14f
Expansive clay, 59–60
can be expensive clay, 2
chemical stabilization with lime, 69, 70f
clue to, 34–35
consolidation test, 59, 59f
crystal structure in control, 63, 64f
field evidence for, 64–65, 64f
guide to, 63
how a layer of expansive clay can cause trouble, 60
managing, 65–67
chainsaw method, 65
observations of strange field behavior, 67, 67f
stripping off the active layer, 65, 66f
structural slabs, grade beams, and piles, 65, 66f
minerals, 5
nature's color coding, 60
new rule for control of, 68, 68f
size grades, defining, 39
transition to, 14
type G, 60
type P, 60
uplift from, 179
why clay expansion stops at 3 layers, 69
Expansive crystal structure, 32–34, 33f

F

Factor of safety, 96, 104, 171, 184
Fall cone test, 27–28, 28f
Fibers to stabilize soil, 118
First-order rate equation (FORE), 50–51, 51f–52f
Flight auger, 18, 126, 182
Flocculated, 39
Floodplain clays, sorting out, 61
what makes river floodplains wide, 61
Floodplain soils of meandering rivers, 62
depth and shape of an oxbow, 62
oxbow lake clay, 62
slack-water or backswamp deposits, 62
Floods
conquering the 100-year flood, 62
Flow net, 76
Fly ash, 197
FORE (first-order rate equation), 50–51, 51f–52f
Foundation engineering. *See* Geology and foundation engineering
Foundations, 2–3
deep. *See* Deep foundations, design of
geology and. *See* Geology and foundation engineering
settlement. *See* Settlement, foundation
shallow. *See* Shallow foundation bearing capacity
Fox, Nathaniel, 197–198
Fracking (hydraulic fracturing), 22, 145–146
Franki, Edgard, 182
Franki piles, 182
Freezing, ground, 197
Friction, 93–96
Amontons' second law, 94–95
angle, 94, 95f
Coulomb's equation, 95–96
Greek connection, 95
ratio, 142–143, 143, 143f
Friction foundations, 8
Friction reversal hypothesis, 200–201, 200f
conditioning, 201
and overconsolidation, 200f, 201
Fröhlich, O.K., 86, 87f
Frost heave and footing depth, 120–122, 120f
Arctic permafrost, 121
elongated lakes, 121
methane release, 121–122
polygonal ground, 121
some practical consequences, 121

G

Gaussian distribution, 35, 35f
Geogrid, 193

Geology and foundation engineering, 1–24
agricultural soil maps, 9
soil series, 9
alluvial soils, distinguishing between, 9–13, 10f–11f
air photo interpretation, 10f–11f, 12–13
alluvial fans, 12
meanders and cutoffs, 10–11, 10f
natural levees, 12
oxbow lake clay, 10f, 11–12
rivers and continental glaciation, 10
slack-water (backswamp) floodplain deposits, 12
earthquakes. *See* Earthquakes
groundwater table(s) influence from, 6–7
groundwater table and soil color, 6, 7t
perched groundwater table, 6–7
intermittent recycling, 7
landslides
landslide scarps, 14–15, 15f
no! no! landslide repair method, 15–16, 15f
not a good place for a patio, 16
recognizing, 16
stopping, 15f, 16–18
overview, 1–2, 21–23
three most common construction materials, 1–2
residual soils, 3–4
travel is wearing, 3–4, 4f
rock that isn't there, 18–21
abandoned mine shafts and tunnels, 20
locating underground caverns, 20
near-surface features, 18, 19f
shallow caverns and sinks, 19, 19f
tunneling machines and rock that isn't there, 20–21
soil layers created by weathering, 4–5
shrinkage cracks and blocky structure in expansive clays, 5, 5f
subsoil "B horizon," 5, 5f
topsoil "A horizon," 4, 5f
soil responses to earthquakes, 21–22
soil types and foundations, 7–9
influence of a groundwater table, 8–9
pull-up of deep foundations by expansive clay, 9
vertical mixing in expansive clay, 6
the walkabout, 23
wind-deposited soils, 13–14
eolian silt deposits, 13–14, 14f
sand dunes, 13
Geophones, 20
Geopier, 197
"Glory holes," 3
Goodman, Jack, 176
Gow, Charles R., 125
Gravel/sand
size grades, defining, 38
Ground freezing, 197
Ground improvement, 3, 193–204
compaction, 194–195
blasting, 195
deep dynamic compaction (DDC), 194
side effects from compaction, 195, 195f
vibratory compaction, 194
friction reversal hypothesis, 200–201, 200f
conditioning, 201
friction reversal and overconsolidation, 200f, 201
grout materials, 197
grout "take," 197
low-slump concrete piers, 203
Mohr's theory, application of, 201–203, 201f
is excavation permitted close to RAPS?, 203
lateral stress and settlement, 202–203, 202f
overview, 193
preloading, 193–194
complex system, 194
enhancing and monitoring the rate of settlement, 193–194
rammed aggregate piers (RAPs), 197–198, 198f
as anchor piers, 203
"saw-tooth" stress pattern, 199
temporary liquefaction, 199, 199f
tension cracks outside the liquefied zone, 199–200
sand piers, 203
soil replacement or improvement, 195–197
stone columns, aggregate, and mixed-in-place piers, 195–197, 196f
when soil does not hold an open boring, 203
Ground-penetrating radar, 20
Groundwater
soil penetration tests (SPTs), 127
table(s) influence from, 6–7
groundwater table and soil color, 6, 7t
perched groundwater table, 6–7
Grout materials, 197
Grout "take," 197
Guyots, 21

H

Hammer blows, analyzing, 173–175
measuring setup with PDA and restrike, 174–175, 175f
pile driving analyzer (PDA), 173–174
wave equation for driven piles, 173, 174f
Hammers
automatic hammer, 131f, 132t
soil penetration tests (SPTs), 130–132
trip hammer, 130
Hands-on experience, 25
Hands-on test for expansive clay, 34
field test, 34
Hansen, Brinch, 117
Hard driving and brooming of wood piles, 170
H-beam piles, 184
Helical screw-in micropiles, 184
High Strain Dynamic Test, 174
Histogram, 35f
Hogentogler, Chester, 29
Hooke's Law, 79f, 125
Housel, W.S., 84
Hutchings, Ian, 93
Hydrogen bonds, 31–32
Hysteresis, 68

I

Independent variable, 104
Injection grouting, 196–197
Intermediate principal stress, 103–104
Intermittent recycling, 7
Inuit igloo, 41

J

Jáky equation, 188
Jet grouting, 184, 196, 196f
Jumping jacks, 194

K

Kansaii International Airport (Kobe, Japan), 52
Kaolinite, 60
crystal structure, 31, 32f
Karst topography, 19, 19f
Katti, R.K., 67, 78
Katti's replacement method, 66f
Köhler, F., 157
K_o Stepped Blade, 161–164, 162f, 166f, 198, 199f
example, 163–164, 164f
interpretation, 163, 164t
test sequence, 163
two-chambered pressure cell, 162–163

L

Lambe, T. William, 103, 315
Lambe's stress path method, 54–55, 103
Lamé, Gabriel, 160
Landslides
landslide scarps, 14–15, 15f
no! no! landslide repair method, 15–16, 15f
not a good place for a patio, 16
recognizing, 16
stopping, 15f, 16–18
chemical stabilization, 17
drainage, 16
drilled quicklime, 17–18
structural restraints: piles, stone columns, and retaining walls, 17
Lateral stress, 157–166
interpretation of pressuremeter test data, 159–161
lateral in situ stress, 159
limit pressure in foundation engineering, 159–160, 159f
soil identifications, 161, 161t
theoretical approach, 160–161, 160f
use in design, 161
K_o Stepped Blade, 161–164, 162f, 166f
example, 163–164, 164f
interpretation, 163, 164t
test sequence, 163
two-chambered pressure cell, 162–163
K values in sand, 187–188, 188t
lower cost, more convenient, 157
overview, 157, 164–165
pressuremeter, 157–159, 158f
self-boring, 158–159, 158f
soil disturbance from drilling, 157
settlement and, 202–203, 202f
Lateral yielding to predict vertical settlement, 153–154, 153f
aging, 153, 154f
Dilatometer shift in direction of major principal stress, 154
Layered crystal structure, 31–32, 33f
Layer thickness, 44
Le Chatelier, 63
Lime, 3, 17
Limit pressure in foundation engineering, 159–160, 159f
Liquefaction, 22, 73–74
definition of, 73, 199
earthquakes, volcanoes, and the "ring of fire," 74
identifying vulnerable soils, 73

Probing with cone penetration tests and the Marchetti Dilatometer (*Cont.*):
cone test data and settlement of foundations on sand, 148
cone test data example, 146–147, 146f
cone tests and foundations on saturated, compressible clay, 148–149
fracking (hydraulic fracturing), 145–146
friction ratio, 142–143, 143f
how can lateral yielding predict vertical settlement?, 153–154, 153f
aging, 153, 154f
dilatometer shift in direction of major principal stress, 154
Marchetti Dilatometer, 150–152, 150f
preparation for testing, 151–152, 151f
soil identification, 152
mechanical versus electrical cones
cone and SPT advantages and disadvantages, 145
piezocone, 144–145, 144f
piezocone and groundwater table, 145
normalizing cone test data for test depth, 147–148
dealing with dimensions, 147–148, 147f
overview, 141
precaution with empirical relationships, 149
predicting settlement, 152, 152f
pushing versus driving, 142, 143f
supplemental cone test data, 149–150
time-outs for pore pressure dissipation, 149
Proctor, R.R., 109
Proctor density test, 109
Progressive nature of a bearing capacity failure, 86, 87f
Pushing versus driving, 142, 143f

Q

Quasi-elastic settlement, 8
Quick clays, 72, 73f
vane shear does not just measure soil cohesion, 72, 73f
Quicksand, 21–22, 76

R

Rammed Aggregate Piers (RAPs), 159, 196, 196f, 197–198, 198f
as anchor piers, 203
"saw-tooth" stress pattern, 199
temporary liquefaction, 199, 199f
tension cracks outside the liquefied zone, 199–200
R^2 analysis, 104
Rankine, W.J.M., 109, 124f
Real factor of safety, 115, 116f
Reese, Lyman, 178
Reference time for 90 percent primary consolidation, 50
Refraction surveys, 20
Reinforced earth, 17, 117–119, 118f, 193
Replacement method, 67–69
between-layer stacking of water molecules, 68–69
hypothesis, 69
new rule for control of expansive clay, 68, 68f
why clay expansion stops at 3 layers, 69
Repose, angle of, 13
Residual soils, 3–4
travel is wearing, 3–4, 4f
Reverse spiral, 183
Reynolds, Osborne, 96
Reynolds number, 96
"Ring of fire," 74
Rivers and continental glaciation, 10
Robbins, James, 20
Rock, 1
end bearing on, 175–176
quality, 176
that isn't there, 18–21
abandoned mine shafts and tunnels, 20
locating underground caverns, 20
near-surface features, 18, 19f
shallow caverns and sinks, 19, 19f
tunneling machines and rock that isn't there, 20–21
Rock Quality Designation (RQD), 176
Rock sockets, 176, 177f

S

Safety hammer, 130, 131f, 132t
Salt versus fresh water clay deposits
size grades, defining, 39
Sample disturbance, correcting for, 53–54, 54f
Sand, 3–4
bearing capacity based on an estimated friction angle, 135
end bearing on, 176–178, 186–187, 187t
size grades, defining, 38 38f
soil penetration tests (SPTs), 128–130, 129t, 132–133
sources of, 13
Sand drains, 108, 193
Sand dunes, 13

Index

Size grades, defining, 38–39
  clay and silt, 39
  expansive versus non-expansive clay, 39
  gravel/sand, 38
  salt versus fresh water clay deposits, 39
  sand/silt, 38, 38f
Skempton, A.W., 72
Skin friction, 2
  and adhesion, 178–179
    depth and differential movement, 178, 179f
    end bearing and skin "friction," 179
    negative skin friction (adhesion), 179, 179f
    uplift from expansive clay, 179
Slab-on-grade concrete slabs, 7
Slack-water (backswamp) floodplain deposits, 12, 62
Sleeve resistance ratio, 143
Slices method, 17
"Slickensides," 64, 65f
Sliding friction, adhesion theory for, 43
Smectites, 34, 63, 69
Smith, E.A.L., 173
Sodium replacing calcium, 34
Soil, 1, 7, 9
  behaving badly. *See* Soils behaving badly
  characteristics related to grain size distribution curves, 38, 39f
  city planning and, 167–168
  classifying. *See* Soil classification
  color and groundwater table, 6, 7t
  consolidation test, 42, 42f–43f
  creep, 44, 108
  earthquakes, responses to, 21–22
  freezing depths in the United States, 120f
  identification, 161, 161t
  layers created by weathering, 4–5
    shrinkage cracks and blocky structure in expansive clays, 5, 5f
    subsoil "B horizon," 5, 5f
    topsoil "A horizon," 4, 5f
  maps, 9, 10f
  profile, 60
  replacement or improvement, 195–197
    stone columns, aggregate, and mixed-in-place piers, 195–197, 196f
  series, 9
  shear strength, evaluating. *See* Soil shear strength, evaluating
  stresses in. *See* Stresses in soils
Soil classification, 25–40
  Atterberg limits and criteria for expansion, 31, 32f
Soil classification (*Cont.*):
  Atterberg limits in soil classification, 29–30, 29f
  clay materials, kinds of, 31–34
    drilling mud, 34
    expansive crystal structure, 32–34, 33f
    going tribal, 34
    layered crystal structure, 31–32, 33f
    when sodium (Na^+) replaces calcium (Ca^{++}), 34
  engineered soil moisture content, 25–26
  expansive clay, clues to, 34–35
  hands-on experience, 25
  overview, 25
  particle sizes, measuring, 35–36, 35f
    defining clay size, 36
    statistical interpretation, 35–36, 35f
  particle sizes determined from sedimentation rates in water, 36–38, 37f
    defining clay size, 38
    performing a sedimentation test, 36–37
  from plastic and remoldable to liquid and flowable, 27
    fall cone test, 27–28, 28f
    plasticity index, 28
    standardizing the LL test, 27, 27f
  size grades, defining, 38–39
    clay and silt, 39
    expansive versus non-expansive clay, 39
    gravel/sand, 38
    salt versus fresh water clay deposits, 39
    sand/silt, 38 38f
  soil characteristics related to grain size distribution curves, 38, 39f
  World War II and new rules for soil classification, 30, 30t, 31f
Soil penetration tests (SPTs), 125–133, 126f
  adjusting the *N* value, 130, 131f, 132–133, 132t, 133f–134f
  groundwater, 127
  hammers, 130–132
  "pocket penetrometer," 128
  sample disturbance, 127–128
  in sand, 128–130, 129t, 132–133
  selective test depths, 127
  Shelby tube samples, 128
  soil mechanics of, 130
Soils behaving badly, 59–78
  braided rivers, 61
  collapsible soils, 70–71
    collapsible alluvium, 71
    delayed collapse, 70–71

Soils behaving badly (*Cont.*):
computers, 76
deep tropical weathering and expansive clay, 63
earthquake dynamics, 75–76, 75f
recurrence intervals, 75–76
expansive clays, 59–60
consolidation test, 59, 59f
crystal structure in control, 63, 64f
field evidence for, 64–65, 64f
guide to, 63
how a layer of expansive clay can cause trouble, 60
managing, 65–67, 69, 70f
nature's color coding, 60
type G, 60
type P, 60
floodplain clays, sorting out, 61
what makes river floodplains wide, 61
floodplain soils of meandering rivers, 61, 62
depth and shape of an oxbow, 62
oxbow lake clay, 62
slack-water or backswamp deposits, 62
liquefaction, 73–74
earthquakes, volcanoes, and the "ring of fire," 74
identifying vulnerable soils, 73
made earthquakes, 73–74
pretreatment to prevent, 74, 74f
overview, 59
quick clays, 72, 73f
vane shear does not just measure soil cohesion, 72, 73f
quicksand, 76
replacement method, 67–69
between-layer stacking of water molecules, 68–69
hypothesis, 69
new rule for control of expansive clay, 68, 68f
why clay expansion stops at 3 layers, 69
wind-deposited soils, regional changes in properties of, 71–72, 71f–72f
Soil shear strength, evaluating, 93–106
bearing capacity and settlement, 93, 94f
a difficult problem, 102–104, 103f
intermediate principal stress, 103–104
Lambe's stress path method, 103
stage testing, 103
direct shear test, 96–100, 97f
Borehole Shear Test (BST), 98–100, 98f–99f
influence of layering, 97–98
friction, 93–96
Soil shear strength, evaluating (*Cont.*):
Amontons' second law, 94–95
Coulomb's equation, 95–96
friction angle and slope angle, 94, 95f
Greek connection, 95
friction angle in soils, 94
dilatancy in design, 96
Mohr's theory, 100–102, 101f–102f
pore water pressure, 102
overview, 93
statistical analysis of test data, 104
R^2 analysis, 104
triaxial shear tests, 104
unconfined compression test, 100, 100f
Spangler, M.G., 88
"Split spoon," 125, 126f–127f, 145
St. Louis Arch, 41
Stage testing, 103
Standard penetration test (SPT) in foundation engineering, 125–140
bearing capacity of sand based on an estimated friction angle, 135
comparisons with measured settlements, 135–136
foundation bearing capacities on clay based on SPT or unconfined compressive strength, 136–137, 138f
net bearing pressure, 137, 138f
reducing settlement with a mat foundation, 137
theoretical foundation design on clay based on unconfined compressive strength, 137
overview, 125, 138
pressure bulb correction, 133–134, 134f
soil penetration tests, 125–133, 126f
adjusting the *N* value, 130, 131f, 132–133, 132t, 133f–134f
groundwater, 127
hammers, 130–132
"pocket penetrometer," 128
sample disturbance, 127–128
in sand, 128–130, 129t, 132–133
selective test depths, 127
Shelby tube samples, 128
soil mechanics of, 130
Steel, 1
Steel pipe piles, 184
Stokes, George, 36
Stokes' Law, 36, 37f
Stone columns, 195, 196f
Stoping, 20, 119
Straw, 118

Strength gains
from aging, 89–90
and slow loading, 171
Stress, lateral. *See* Lateral stress
Stresses in soils, 79–91
adapting the Boussinesq theory, 80–81, 80f
approximating pressure distributions, 81–82, 82f
concentrated stresses, 79, 79f
convenient maximum depth for pressure calculations, 90
foundation stresses transferred to nearby unyielding walls, 88–89, 89f
overview, 79
plate bearing test
as a model foundation, 84–85, 84f
performing, 85–86, 86f
on weathered soil profiles, 86–87, 88f
preloading, 82–84
progressive nature of a bearing capacity failure, 86, 87f
a snag in the relationship, 81, 81f
strength gains from aging, 89–90
interruptions during pile driving, 90
Structural restraints, 17
Structural slabs, 65, 66f
Subsoil "B horizon," 5, 5f

T

Tapered piles, 184
Temporary liquefaction, 199, 199f
Tension breaks in concrete piles caused by pile driving?, 170
piles doing a U-turn, 170
Tension cracks outside the liquefied zone, 199–200
Terzaghi, Karl, 29, 41–42, 43f, 47, 126, 132–133, 135, 169
Terzaghi approach, 111–112, 112f, 113t
Terzaghi bearing capacity factors, 112–115, 113t, 114f
alternative solutions, 115
local shear, 115
Thixotropy, 16, 34
definition of, 27
Tiebacks, 193
Time and foundation settlement, 43–44, 44f
Time-outs for pore pressure dissipation, 149
Topsoil "A horizon," 4, 5f
Tremie, 180
Triaxial shear test, 100, 101f, 104
Trip hammer, 130
Tunneling machines and rock that isn't there, 20–21
Two-chambered pressure cell, 162–163
Type G expansive clays, 60
Type P expansive clays, 60

U

Unconfined compression test, 100, 100f
Underconsolidated soil, 8
"Undisturbed," 1
Unified System, 30, 30t
Unsaturated soil mechanics, 67
Uplift from expansive clay, 179
U.S. Department of Transportation, Federal Highway Administration (DOT-FHWA), 161, 198

V

Vane shear test, 72, 73f
Varved clay soil, 149
Vertical mineshafts, dangers of, 119
Vertical mixing in expansive clay, 6
Vertisol, 6, 65, 66f
Vesic, Aleksandar, 116, 169
Vibratory compaction, 194
Vibratory pile drivers
avoid impacts and are less noisy, 175
Vibroflot, 195
Vidal, Henri, 118
Void ratio, 42, 45, 46t
calculating, 45, 46t
Volcanoes, 21
Vulnerable soils, identifying, 73

W

Wall foundations, 7–8
Water molecules, between-layer stacking of, 68–69
Wave equation for driven piles, 173, 174f
Wegener, Alfred, 21, 24f
Well graded, 38
Wellington, A.M., 170
Wick drains, 194
Wind-deposited soils, 13–14
eolian silt deposits, 13–14, 14f
regional changes in properties of, 71–72, 71f–72f
sand dunes, 13
Wood piles, 169
World War II and new rules for soil classification, 30, 30t, 31f